Assessing the Effectiveness of Future Concepts in the U.S. Air Force

Application to Future Logistics Concepts

DON SNYDER, KRISTIN F. LYNCH, ALEXIS A. BLANC,
JONATHAN L. BROSMER, JOHN G. DREW, KYLE HAAK,
MYRON HURA, DANIEL ISH, KELLY KLIMA, FABIAN VILLALOBOS

Prepared for the Department of the Air Force
Approved for public release; distribution unlimited

 PROJECT AIR FORCE

For more information on this publication, visit **www.rand.org/t/RRA534-1**.

About RAND

The RAND Corporation is a research organization that develops solutions to public policy challenges to help make communities throughout the world safer and more secure, healthier and more prosperous. RAND is nonprofit, nonpartisan, and committed to the public interest. To learn more about RAND, visit www.rand.org.

Research Integrity

Our mission to help improve policy and decisionmaking through research and analysis is enabled through our core values of quality and objectivity and our unwavering commitment to the highest level of integrity and ethical behavior. To help ensure our research and analysis are rigorous, objective, and nonpartisan, we subject our research publications to a robust and exacting quality-assurance process; avoid both the appearance and reality of financial and other conflicts of interest through staff training, project screening, and a policy of mandatory disclosure; and pursue transparency in our research engagements through our commitment to the open publication of our research findings and recommendations, disclosure of the source of funding of published research, and policies to ensure intellectual independence. For more information, visit www.rand.org/about/principles.

RAND's publications do not necessarily reflect the opinions of its research clients and sponsors.

Published by the RAND Corporation, Santa Monica, Calif.
© 2022 RAND Corporation
RAND® is a registered trademark.

Library of Congress Cataloging-in-Publication Data is available for this publication.
ISBN: 978-1-9774-0867-9

Cover: Valerie Seelye/U.S. Air Force.

About This Report

This report presents a Future Logistics Concept Assessment Framework, which is a disciplined, systematic way to assess proposed future logistics concepts to meet the requirements of the National Defense Strategy. The framework prompts the questions that decisionmakers need to answer to know whether to pursue such a future concept. This report should be of interest to the logistics community in the U.S. Air Force but has wider application to any organization proposing a new concept for operations.

The research reported here was commissioned by the Director of Resource Integration, Deputy Chief of Staff for Logistics, Engineering and Force Protection and conducted within the Resource Management Program of RAND Project AIR FORCE as part of a fiscal year 2020 project titled "Contested Logistics."

RAND Project AIR FORCE

RAND Project AIR FORCE (PAF), a division of the RAND Corporation, is the Department of the Air Force's (DAF's) federally funded research and development center for studies and analyses, supporting both the United States Air Force and the United States Space Force. PAF provides the DAF with independent analyses of policy alternatives affecting the development, employment, combat readiness, and support of current and future air, space, and cyber forces. Research is conducted in four programs: Strategy and Doctrine; Force Modernization and Employment; Resource Management; and Workforce, Development, and Health. The research reported here was prepared under contract FA7014-16-D-1000.

Additional information about PAF is available on our website:
www.rand.org/paf

This report documents work originally shared with the DAF on November 16, 2020. The draft report, issued on September 18, 2020, was reviewed by formal peer reviewers and DAF subject-matter experts.

Acknowledgments

We thank Edwin Oshiba for sponsoring this project and for his support throughout. Colonel Margaret (Maggie) Sleeper was a supportive action officer. We also appreciate the support we received from the Air Force Expeditionary Center, EC-A3/4/5.

At RAND, we thank Paul Dreyer, Brian Killough, Chad Manske, Sarah MacConduibh, and Patrick Mills for comments on and suggestions to the project. Formal reviews by Jim Leftwich and Michael Vermeer improved the report.

That we received help and insights from those acknowledged above should not be taken to imply that they concur with the views expressed in this report. We alone are responsible for the content, including any errors or oversights.

Contents

About This Report.. iii

Figures and Tables vi

Summary vii

1. Future Challenges for Logistics 1
   The Problem Setting 2
   Terminology 4
   Objectives and Research Approach 4
   Efficiency Versus Robustness 7
   Analysis, Judgment, and Decisions 9
   Demonstrating the Framework 10

2. Assessing Versatility in Terms of Benefits and Downsides 11
   Defining the Operational Versatility of Future Logistics Concepts 11
   Changing Geopolitical Setting 13
   Threat Environment 19
   Evolving Blue Operational Concepts of Employment 21
   Summary 24

3. Assessing Implementation Challenges 26
   Doctrine 27
   Organization 27
   Training 29
   Materiel 29
   Leadership and Education 31
   Personnel 31
   Facilities 31
   Policy 31

4. Case Study of the Assessment Processes 33
   Purpose of the Case Study 33
   Example Case Study 33

5. Afterword 44
   Applicability of the Framework 44
   Target Users of the Framework 44
   Utility of the Framework 45

Appendix. Future Logistics Concept Assessment Framework Checklist 46

Abbreviations 50

References 51

Figures and Tables

Figures

Figure S.1. Research Approach viii

Figure 1.1. Approach for the Framework 7

Figure 2.1. Schematic Depiction of Dimensional Factors for Assessing Change to
Versatility 25

Tables

Table 2.1. Summary of Operational Versatility Dimensions 24

Table 3.1. Notional Logistics Implementation Equities 28

Summary

Issue

The National Defense Strategy (NDS) describes a stressing set of future challenges for U.S. Air Force logistics.[1] To meet these challenges, the U.S. Air Force will entertain novel ideas—what we call *future logistics concepts*—to adapt the way that it organizes, trains, and equips for logistics to support the NDS.[2] Future logistics concepts might be materiel or nonmateriel solutions, and they will need to be assessed for overall utility and be put through an analytical gauntlet to expose any weaknesses so that the U.S. Air Force is well prepared to defend a concept to the Office of the Secretary of Defense and Congress.

Rather than recommending a specific future logistics concept, we present a Future Logistics Concept Assessment Framework, which is a disciplined, systematic way to assess proposed future logistics concepts. The framework asks the questions that decisionmakers need to answer to know whether to pursue a proposed option. It is designed to be easily executable by an action officer without extensive expertise. Results should inform decisions by revealing benefits and downsides and should not advocate for or against a future logistics concept.

Approach

The assessment presents the following five steps, which are shown in Figure S.1:

- Identify the operational benefits sought by the future logistics concept.
- Define the future logistics concept.
- Assess: How versatile is the future logistics concept for meeting future operational needs?
- Assess: What are the implementation challenges for transitioning from the status quo to the future logistics concept?
- Present well-reasoned assessments—not recommendations—to senior decisionmakers.

---

[1] U.S. Department of Defense, *Summary of the 2018 National Defense Strategy of the United States of America: Sharpening the American Military's Competitive Edge*, Washington, D.C., 2018, p. 2.

[2] Charles Q. Brown, Jr., *Accelerate Change or Lose*, Washington, D.C.: U.S. Air Force, August 2020.

Figure S.1. Research Approach

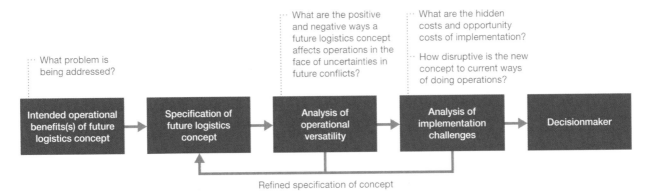

Results

The core of the Future Logistics Concept Assessment Framework lies in the two assessment steps of operational versatility and implementation challenges.

Operational Versatility

Rather than assessing the operational effectiveness of a future logistics concept against one or more scenarios, operational versatility assesses how well the concept adapts to a variety of possible futures. It does so in the following three dimensions:

- the concept's versatility against a changing geopolitical setting
- the concept's versatility against the variety of future threats
- the concept's versatility to support future operational concepts of employment.

These dimensions are discussed in greater detail, with a worked case study for illustration, in this report.

Implementation Challenges

To assess these challenges, we organize the necessary changes to the status quo to implement the future logistics concept by changes to doctrine, organization, training, materiel, leadership and education, personnel, facilities, and policy (DOTMLPF-P), although for the purpose of future logistics concepts, we address issues that go beyond joint guidance. Each of the DOTMLPF-P dimensions is discussed in greater detail, with a worked case study for illustration, in this report.

1. Future Challenges for Logistics

> The stress on quantifiable elements is particularly risky in cost-benefit work where objectives are hard to define or subject to change. In most cases the cost elements can be reduced to money terms. By contrast, objectives may be numerous, mutually incommensurable, and reducible to money terms only on the basis of rather arbitrary and subjective judgments by the analyst.
>
> — James R. Schlesinger[3]

> Simply stated, the purpose of an analysis is to provide illumination and visibility—to expose some problem in terms that are as simple as possible. This exposé is used as one of a number of inputs by the decisionmaker. Contrary to popular practice, the primary output of an analysis should not be conclusions and recommendations.
>
> — Glenn A. Kent[4]

As the future warfighting environment evolves, numerous proposals will be advanced for how U.S. Air Force logistics can meet new challenges. Some of these proposals will emerge from within the U.S. Air Force by innovative thinkers. Others will come from advocates outside the U.S. Air Force, including some who are motivated to sell a product or service. In these situations, there is generally no shortage of advocacy for the rich benefits the new idea allegedly will bring. But many ideas also bring downsides, some of which might be evident only with careful analysis. And many ideas also come with implementation challenges that stress the whole enterprise. This report presents a disciplined, systematic way to assess proposed future logistics concepts. We describe a framework that asks the questions that decisionmakers need to answer in order to know whether to pursue a proposed option. We call this a *Future Logistics Concept Assessment Framework*.

The U.S. Air Force is embarking on a period of change.[5] Adopted changes need to be sound and defensible. The methods in this report should help the U.S. Air Force avoid regrets from pursuing future logistics concepts that eventually disappoint. But even sound ideas can falter if their advocates are not prepared to defend them. Therefore, this method is also useful in running novel ideas through an analytical gauntlet to expose any weaknesses so that, if the U.S. Air Force chooses the option, it will be well prepared to defend it to the Office of the Secretary of Defense and Congress.

---

[3] James R. Schlesinger, *Systems Analysis and the Political Process*, Santa Monica, Calif.: RAND Corporation, P-3464, 1967, pp. 3–4.

[4] Glenn A. Kent, with David Ochmanek, Michael Spirtas, and Bruce R. Pirnie, *Thinking About America's Defense: An Analytical Memoir*, Santa Monica, Calif.: RAND Corporation, OP-223-AF, 2008, p. 94.

[5] Charles Q. Brown, Jr., *Accelerate Change or Lose*, Washington, D.C.: U.S. Air Force, August 2020.

The Future Logistics Concept Assessment Framework is designed for use early in the consideration of future concepts. The idea is to reveal the most-promising concepts and to prune the least promising before significant investment of resources. The arguments the framework produces also will be useful for justifying and defending budget positions once the concept is adopted.

The Problem Setting

In war, logistics can be decisive. Operational forces need copious amounts of ammunition, food, fuel, repair parts, and a wide variety of other support. Military history is replete with examples of forces besting their adversaries despite weaker numbers and being more poorly equipped because they were better able to move necessary materiel to the warfighter at the right time and place. Success often hinged on the fulcrum of logistics.[6]

Because logistics can be decisive, operations can be placed at risk by the failure of logistics. Logistics can fail for many reasons. An adversary might directly target logistics, military forces on one side might be more heavily dependent on logistics than the forces on the other side, or the force structure for logistics might not be well suited to the conflict it needs to support. All three of these areas could be pitfalls for future U.S. Air Force logistics.

The future outlined in the National Defense Strategy (NDS) is a stressing one for U.S. Air Force logistics. The NDS describes a "reemergence of long-term, strategic competition" with China and Russia and the need to prepare for potential conflict with North Korea or Iran that could involve nuclear, chemical, or biological warfare.[7] Some of the challenges this situation poses to U.S. Air Force logistics are

- expeditionary operations with geographically extended logistics supply lines potentially operating for a protracted duration
- operations against adversaries with near-peer or peer attack capabilities against logistics
- the ability to prosecute logistics operations in a nuclear, chemical, or biological environment
- the possibility of attacks on the U.S. homeland, including commercial logistics support entities, such that no place or logistics element is a sanctuary.

To address these challenges, the NDS directs the services to organize, train, and equip for " . . . ground, air, sea, and space forces that can deploy, survive, operate, maneuver, and regenerate in all domains while under attack. Transitioning from large, centralized, unhardened infrastructure to smaller, dispersed, resilient, adaptive basing that include active and passive

---

[6] See, for example, Jobie Turner, *Feeding Victory: Innovative Military Logistics from Lake George to Khe Sanh*, Lawrence, Kan.: University Press of Kansas, 2020.

[7] U.S. Department of Defense (DoD), *Summary of the 2018 National Defense Strategy of the United States of America: Sharpening the American Military's Competitive Edge*, Washington, D.C., 2018, p. 2.

defenses will also be prioritized."[8] The NDS adds that "[i]nvestments will prioritize prepositioned forward stocks and munitions, strategic mobility assets, partner and allied support, as well as non-commercially dependent distributed logistics and maintenance to ensure logistics sustainment while under persistent multi-domain attack."[9]

This direction adds further challenges for U.S. Air Force logistics:

- the need to operate out of numerous, small sites
- the ability to support operational ground maneuver of air forces.

To meet these challenges, the U.S. Air Force will need to entertain a variety of ideas to adapt the way that it organizes, trains, and equips for logistics. We call these proposed changes *future logistics concepts*—i.e., novel ideas to adapt the way that the U.S. Air Force organizes, trains, and equips for logistics to support the NDS. This term overlaps with the term *capability development*, which is used in the Air Force Warfighting Integration Capability. Although the Air Force Warfighting Integration Capability is one target community for which we developed this framework, we expect that others will find this framework useful. Some of the ideas that other organizations might address could extend beyond capability developments.

Future logistics concepts can be materiel or nonmateriel changes. They might be specific materiel additions to address a single issue, such as deployable nuclear power plants,[10] or new deployment ideas, such as moving cargo via rockets through space.[11] Future logistics concepts also might be novel ways to manage processes, such as distributed ledger (blockchain) technologies,[12] or broader logistics concepts for supporting new employment concepts, such as changes to combat service support to enable dispersed operations.[13] Or, they might be whole-scale changes to processes, command and control, or any other idea for improving overall logistics.

---

[8] DoD, 2018, p. 6.

[9] DoD, 2018, p. 7.

[10] Tony Bertuca, "DOD Awards Contracts for Prototype Mobile Nuclear Reactor," *Inside Defense*, March 9, 2020; Defense Science Board, *Task Force on Energy Systems for Forward/Remote Operating Bases: Final Report*, Washington, D.C.: U.S. Department of Defense, Office of the Under Secretary of Defense for Acquisition, Technology, and Logistics, August 1, 2016; Steve Trimble, "Nuclear Air Force?" *Aviation Week & Space Technology*, November 25–December 8, 2019.

[11] Sandra Erwin, "Space Launch Vehicles Eyed by the Military to Move Supplies Around the World," *Space News*, August 2, 2018; Joe Pappalardo, "The Air Force Is Actually Considering Rocket Launches to Move Cargo Around the Globe," *Popular Mechanics*, October 29, 2018.

[12] Chris Baraniuk, "Blockchain: The Revolution That Hasn't Quite Happened," BBC News, February 11, 2020; Shaun Waterman, "Air Force Cyber Weapons Factory Moves Ahead with Blockchain Project," *Air Force Magazine*, July 17, 2020; Roger Wattenhofer, *Distributed Ledger Technology: The Science of the Blockchain*, 2nd ed., San Bernardino, Calif.: Inverted Forest Publishing, 2017.

[13] Kyle Mizokami, "In a Future War, the Air Force's Big Air Bases Could Be a Big Liability," *Popular Mechanics*, February 13, 2020.

Terminology

The U.S. Air Force sometimes uses different terms than its joint partners. Once such term causing confusion is the term *combat support*. The U.S. Air Force uses *combat support* to mean "[t]he foundational and crosscutting capability to field, base, protect, support, and sustain U.S. Air Force forces across the range of military operations."[14] In contrast, the U.S. Department of Defense (DoD) defines the same term, *combat support*, as "[f]ire support and operational assistance provided to combat elements."[15] The joint term analogous to the U.S. Air Force term *combat support* is *combat service support*, which is defined as "[t]he essential capabilities, functions, activities, and tasks necessary to sustain all elements of all operating forces in theater at all levels of warfare."[16] Both the U.S. Air Force and DoD define logistics as "[p]lanning and executing the movement and support of forces."[17] In both U.S. Air Force and DoD usage, security forces are considered part of combat support. That is, the U.S. Air Force considers security forces to be part of logistics, whereas DoD does not. For this reason, in this report, we avoid the term *combat support* and use the term *logistics*.

Objectives and Research Approach

When addressing future challenges, the more innovative the U.S. Air Force logistics community is, the greater the number and novelty of future logistics concepts that will be conceived and proposed for development and implementation. Regardless of type, the U.S. Air Force will need to evaluate the merits of each concept and prioritize them against the status quo. What operational benefits does each concept bring? What operational downsides does each bear, and do its expected benefits warrant accepting any of those downsides? What are the costs of a future logistics concept in terms of the institutional disruptions for implementation? That is, what would it take to transition from the status quo to the future logistics concept? To better answer these questions, U.S. Air Force decisionmakers need a systematic analysis process to support their decisions and strengthen their advocacy of those decisions to DoD and Congress.

Operational Benefits and Downsides

We take as a point of departure that the exact circumstances that U.S. Air Force combat service support force structure will face in the future is uncertain. The nature and setting of

---

[14] U.S. Air Force, *Air Force Glossary*, Maxwell Air Force Base, Ala.: Curtis E. LeMay Center for Doctrine Development and Education, July 18, 2017. The U.S. Air Force often adds the prefixes *agile* or *expeditionary* to combat support.

[15] Chairman of the Joint Chiefs of Staff, *DOD Dictionary of Military and Associated Terms*, Washington, D.C.: U.S. Department of Defense, January 2020.

[16] Chairman of the Joint Chiefs of Staff, 2020.

[17] Chairman of the Joint Chiefs of Staff, 2020.

conflicts often come as a surprise, differing from the best-laid plans: Which countries will allow overflight and access to operational locations? Which operations will those countries permit? The answers to both of these questions will change with geopolitical circumstances. Exactly how the threat environment will unfold over time is not precisely known. And how the U.S. Air Force structure and Blue concepts of employment (CONEMPs) will evolve depends on these uncertain factors and on technological opportunities.[18] Evaluating operational benefits and downsides should reflect these uncertainties. More weight should be given to future logistics concepts that solve a wide variety of challenges than those that solve narrow problems in particular circumstances.

The hazards of preparing for a singular future are well illustrated by the fall of France in 1940. The lesson that France learned from World War I was the importance of defense and the perils of aggressive offense. Memories of Verdun played heavily in its military planning in the 1930s. Minister of War André Maginot championed the idea of heavily investing military resources in an elaborate line of fixed defensive fortifications. Had warfare in the 1940s resembled that of World War I, France would have been well prepared. But technologies and stratagems evolved, and the warfare of World War II was one of joint operations and maneuver; the advantage went to offense. The French, having postured their forces for a single future that did not happen, fell to German forces in a little more than a month, despite having roughly comparable military forces.[19] The French would have been better off planning for a spectrum of possible futures.

The most promising future logistics concepts will be those that are expected to work over a spectrum of circumstances to address the uncertainties of future needs. No single word quite captures all of the dimensions of this concept—that logistics be sufficiently flexible to support uncertain environments but not break, that logistics be able to absorb and recover from attacks, and that logistics be sufficiently adaptable to support evolving Blue CONEMPs. Furthermore, logistics should be sufficiently timely (or agile) across all of these aspects. In this report, we use the term *versatile* to capture this range of attributes. As we elaborate below, decisionmakers will need to determine how much they value any dimension of versatility or whether they prefer a niche solution to a very specific problem. However, they should be aware of the benefits and downsides of a future logistics concept across the spectrum.

A key part of any assessment of future logistics concepts, therefore, should focus on assessing the versatility a concept renders to logistics. Developing a framework for assessing versatility raises two questions:

---

[18] A *CONEMP* "adds depth and detail to concepts of operations by providing specific solutions and required capabilities required to execute a specific mission, function, or task" (Chairman of the Joint Chiefs of Staff Instruction 3030.01, *Implementing Joint Force Development and Design*, Washington, D.C., December 3, 2019, p. GL-5).

[19] Alistair Horne, *To Lose a Battle: France 1940*, Boston, Mass.: Little, Brown and Company, 1969.

- Relative to what factors should future logistics concepts be versatile?
- How can the versatility of logistics as a general attribute be assessed?

We address these questions in Chapter 2.

Implementation Challenges

Transforming from the status quo to a future logistics concept will present implementation challenges. These challenges come at a cost to the U.S. Air Force. Some future logistics concepts might bring efficiencies and reduce overall costs. Others might bring higher operational versatility at the expense of reduced efficiencies during peacetime. Costs come in many forms. For future logistics concepts that involve materiel changes, procurement and sustainment costs arise. For both materiel and nonmateriel concepts, additional costs could accrue for training and staffing tasks. Some future logistics concepts might introduce enough of a change that they induce a need to alter organizational structure or culture. Changes of this magnitude displace other priorities and therefore introduce opportunity costs. In going from the status quo to the future logistics concept, all of the necessary changes to the U.S. Air Force (and other entities) and the associated costs comprise these implementation challenges. These changes will be an important factor in deciding on a future logistics concept, even if the concept brings considerable operational benefits with acceptable operational downsides. Assessing these implementation challenges is the subject of Chapter 3.

Methodology

Our methodology for developing this framework was to flesh out (1) the challenges presented in the NDS to develop dimensions of operational versatility and (2) the doctrine, organization, training, materiel, leadership and education, personnel, facilities, and policy (DOTMLPF-P) viewpoint to develop dimensions of implementation challenges.

Framework Structure

Figure 1.1 schematically depicts our overall research approach. From left to right, the first box indicates a need to define the problem that the future logistics concept is meant to address. What is the expected operational benefit over the status quo? Problem specification leads to defining the future logistics concept. We separate the step of defining the future logistics concept from that of determining its objectives to emphasize that they are independent activities. The first defines objectives, or *what* is sought. The second defines *how* that objective is expected to be met. The first is aspirational, while the second is a concrete, implementable proposal. The future logistics concept might be purely materiel, nonmateriel, or a combination of the two. The more specificity the definition has, the better it can be assessed.

The next step is assessing the benefits and downsides of a future logistics concept in terms of versatility. To do this, our point of departure is the NDS and the challenges for logistics that it raises. The guiding approach in Chapter 2 for assessing versatility is how much the future logistics concept is likely to advance the ability of U.S. forces to meet NDS objectives. In Chapter 3, the guiding approach for assessing the implementation challenges is organized around the changes germane to DOTMLPF-P. To address future logistics concepts, our discussion of DOTMLPF-P introduces additional aspects that are not specifically mentioned in joint guidance.

Both the assessment of versatility and the costs could very well reveal that the specification of the future logistics concept lacks enough detail to make a fully informed assessment. For example, a future logistics concept might not specify enough detail about how information is being processed, stored, and transmitted to assess how it might change vulnerability to adversarial cyber operations. At this juncture, we recommend that a choice be made either to note that some aspects are not amenable to being assessed until further refinement of the future logistics concept or to refine the concept itself. The option for refinement is depicted by the arrow at the bottom of Figure 1.1.

Figure 1.1. Approach for the Framework

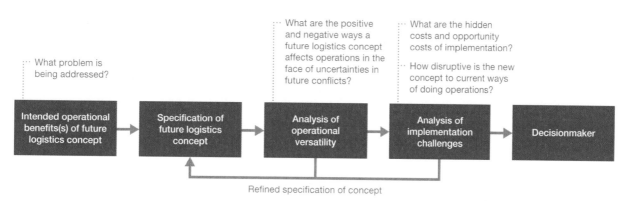

The final step, depicted by the box on the far right of Figure 1.1, is to present the assessments from analysis to decisionmakers. We will take up some considerations of the decisionmaking process and the role of analysis at the close of this chapter. But first, we discuss some issues around lowering costs and increasing operational effectiveness in a high-end fight—specifically, the tension between efficiency and robustness. This tension can affect the emphasis placed on the versatility and implementation challenges.

Efficiency Versus Robustness

A tension naturally exists between operational efficiency (during steady-state operations) and operational robustness (during disturbances to steady-state operations). Achieving high operational versatility and low implementation challenges shares this tension. To some degree, all organizations confront this dilemma. The operation of commercial supply chains is a good example. Market pressures drive the commercial sector to seek efficiencies in the supply chain during normal, steady-state operations. One response is an element of lean logistics, in which the supply stock of parts is minimized and parts arrive at the place of need "just in time." However, a well-tuned just-in-time supply chain that is calibrated for steady-state demand can be fragile under duress. The inability of the commercial supply chain to handle the abrupt change in consumer demand during the early months of the coronavirus disease of 2019 (COVID-19) pandemic is a good example.[20] The commercial sector often positions itself on this trade-off spectrum by assessing the savings during normal operations (efficiencies) and the costs of interrupted operations during times of duress (cost of lack of robustness). It then balances these trade-offs to determine how much to invest in robustness.

DoD experiences this same tension. Conflicting guidance within the government compounds the tension. On one hand, the budget process places topline spending constraints on the U.S. Air Force, often with few programs available for offsets, which forces hard decisions about how to balance the budget. Efficiencies are frequently sought to find offsets, supposedly without loss—or with minimal loss—of capability. On the other hand, strategies and plans, which often are fiscally unconstrained, emphasize operational effectiveness and robustness. Documents like the NDS exhort the services to develop new capabilities for operating in a more challenging environment and ensuring the robustness of operations against near-peer (or peer) attacks.[21] Although robustness and efficiencies often are in tension, there are certainly instances in which an option is both robust and efficient, but this happy circumstance is not the general case.

When seeking a judicious balance between efficiency and robustness, DoD confronts challenges that differ from those of the commercial sector. In the commercial sector, a default state of operations exists that is in some sense normal or a steady state. For an automobile manufacturer, for example, this is the planned production rate. Companies exist to meet these normal operational conditions. External factors that significantly disturb this state, such as natural disasters, wars, and pandemics, are nuisances from which the company needs to recover. Many companies do not need to thrive during the period of duress; they merely need to survive.

In contrast, DoD operations exist for the disturbed state—war.[22] The normal, day-to-day, steady-state operations for organizing, training, and equipping activities during peacetime only exist so that the services can supply forces to meet wartime demands. So long as a company can see its way through some kind of catastrophic shock to operate normally in the future, the company might be willing to accept the risk of significantly disrupted activities during duress (e.g., natural disasters). But DoD must be able to operate with some proficiency during the trials of wartime. Otherwise, any efficiencies accumulated during peacetime are illusory. The money expended for organizing, training, and equipping, no matter how efficient, is wasted if the force cannot perform adequately during wartime. For this reason, we focus on robustness (or, more

---

[20] Lizzie O'Leary, "The Modern Supply Chain Is Snapping," *The Atlantic*, March 19, 2020.

[21] DoD, 2018, p. 4.

[22] More accurately, the military exists to prevent war through deterrence and is prepared for war should deterrence fail. But to be an effective deterrent, the military must be convincingly capable of meeting wartime demands.

generally, what we call versatility) rather than efficiency (in this context, lowering implementation challenges or costs) as the primary need of any future logistics concept.

A second difference is that the commercial sector can express the savings accrued by efficiencies, the costs of robustness, and the penalty for not having sufficient robustness all in the same measure—dollars. To use a term from economics, all three are *commensurable*. For DoD, however, although peacetime efficiencies and the costs of instituting wartime robustness measures are commensurate and can be measured in dollars, the regrets of not fulfilling national objectives during wartime are not so easily measured in monetary terms. For DoD, the three are incommensurate.

Analysis, Judgment, and Decisions

Analysis plays a substantial role in assessing the benefits, downsides, and costs of future logistics concepts. Much of this report offers a way to structure this analysis to inform decisions about future logistics concepts. But it is important to understand the role and limits of analysis in this context.

The tension between efficiency and robustness introduces several nuances about how decisions should be made. The existential need for effective operations during the stress of war means that there is a fundamental priority within DoD for operational effectiveness (during wartime) over efficiency (during peacetime). However, real budgetary constraints and direction from above the service level mean that risks during wartime often will need to be acknowledged and accepted to stay within available peacetime budgets. Analysis reveals and illuminates those risks, but where to accept those risks, and how much risk to accept, are questions for military and political judgment. The fact that these trades are incommensurate limits the ability of analysis to fully quantify them on a comparable basis, again leading to military and political judgment.

Furthermore, although analysis can reveal the nature of risks posed to a specific mission, the various missions are, from the vantage point of analysis, incommensurate. Analysis can show how one investment that supports air superiority balances with another investment that supports that same mission, but it fails to show the relative value of air superiority with, say, nuclear deterrence. Analysis can show whether a future logistics concept contributes narrowly to one mission or widely to a variety of missions. But because these missions are incommensurate, analysis cannot determine whether the nation should accept risk in one mission versus another. Those decisions lie in the domain of military and political judgment.

Therefore, the task of determining which future logistics concepts are promising enough to pursue, and where they fall on the trade-off spectrum of efficiency and robustness, is not an optimization problem. It is a problem of judgment, albeit judgment informed by analysis. The role of the analyst is not to make final recommendations.[23] It is to show the full range of benefits

---

[23] See, for example, Kent, 2008.

and the full range of downsides in the context of the uncertainties of future warfare demands. These findings, along with cost, provide the information from which decisions based on military and political judgment are made.

In addition, these decisions, informed by analysis, will not resemble final and definitive resolutions, but instead resemble continuous evaluations as circumstances evolve. By their very nature, future logistics concepts will range from innovative concepts that are early in their maturity to adaptations of proven concepts from the commercial sector. Also, the environment will change continuously. A future logistics concept that is judged to be unworthy at one time could very well be judged worthy later. The evaluation process, therefore, needs to be continuous.

Simply getting the "right" answer is not enough. It is also imperative that the U.S. Air Force successfully advocate for new concepts within government (and against the legacy concept that would be replaced). To ensure stable funding, the U.S Air Force also will need to anticipate and address counterarguments for any future logistics concept that the Office of the Secretary of Defense or Congress might raise, show that the concept objectively provides the robustness needed during wartime, and show that its costs are affordable. Systematic, objective, and reproduceable analysis is essential to form these arguments. It is the goal of this report to provide a systematic way to inform these decisions, thus providing a counterweight to the rosy presentations of advocates for future logistics concepts. Essentially, this report asks, What questions should a decisionmaker ask of any future logistics concept?

Demonstrating the Framework

In Chapter 4, we demonstrate this approach with a case study of a future logistics concept. The goal of the demonstration of the framework is twofold—to ensure that the process identifies the central issues and to show how the framework can be applied with real, worked examples. The case study is designed to show how to inform decisionmakers about the identity and rough impacts of risks mitigated or accepted by adopting a future logistics concept. We do not advocate for or against any of these notional concepts. Analysis alone cannot arrive at a recommendation. The purpose of analysis is to inform decisionmakers of the risks and benefits of a course of action under consideration. Whether to accept those risks and expend resources to mitigate new risks belongs to the decisionmaker, not the analyst.

2. Assessing Versatility in Terms of Benefits and Downsides

> War is the realm of chance. . . . Chance makes everything more uncertain and interferes with the whole course of events.
>
> — Carl von Clausewitz[24]

The first step in addressing the operational benefits and downsides of future logistics concepts in an uncertain future is to stipulate what a concept might need to address. There is the future world that we expect to happen, and there is the future world that actually unfolds. Although the world we expect to happen and plan for might be one state or a small number of states, the actual future world can be any of an uncountable number of states. To assess versatility, we need to assess against the many possible states, not a singular expected state.

Future logistics concepts, therefore, need to be assessed against a comprehensive range representing the spectrum of possible future environments that might stress logistics. The first task is to define the underlying dimensions that form a basis for these possible future environments. These dimensions are not predictions of the future and are not "scenarios" in the usual sense of the word. They are key ways in which the future might vary and, therefore, are the variables against which versatility is assessed.

There is probably no way to ensure that all of the dimensions can be enumerated. A list of dimensions should be general enough to avoid overly specifying the future, yet be sufficiently comprehensive in scope and detailed enough to reveal pertinent risks to decisionmakers in a useable form. Broadly speaking, three overall dimensions of the future—the changing geopolitical setting, the threat environment, and evolving Blue CONEMPs—drive the six challenges listed in Chapter 1 that emerge from the NDS. Each dimension has several subdimensions that can be defined and assessed independently. Delineating these dimensions and how they bear on versatility is central to the assessment framework. We devote this chapter to that topic.

Defining the Operational Versatility of Future Logistics Concepts

We assess versatility by evaluating how well a future logistics concept performs along each of the three dimensions—the changing geopolitical setting, the threat environment, and evolving Blue CONEMPs. For some dimensions, that performance might be to decrease assessed risk. For other dimensions, the performance might preserve the status quo of risk, or even increase risk. In

---

[24] Carl von Clausewitz, *On War*, edited and translated by Michael Howard and Peter Paret, Princeton, N.J.: Princeton University Press, 1976, Book One, Chapter Three, p. 101.

all cases, the benefits and downsides are those of operations, not logistics performance. Operations should be viewed from the joint perspective to the maximum extent possible.

In Chapter 4, we work through an example along the dimensions described in this chapter, but an example here will help convey the spirit of the analysis. Consider an option—in anticipation of a conflict with a specific potential adversary—of prepositioning equipment in the countries of current allies or partners. The primary purpose of prepositioning is to decrease the time to get access to the equipment in the event of a conflict. Are there potential downsides?

If it is placed close to the adversary, the equipment could become overrun by ground operations or be damaged or destroyed by kinetic attack. If the equipment is placed too far away from where the potential conflict erupts, supply lines would become extended, requiring more time to supply the front lines. If it is placed in a single host nation to avoid aggravating the adversary, that host nation might not allow access to the equipment at the time of need. Spreading the equipment across a variety of nations reduces that risk. Yet the very act of storing equipment in theater could be escalatory, provoking the adversary to take countermeasures in preparation for conflict. In still other dimensions, there would be no change to the status quo of risk. For example, whether the equipment is stored in one place or another might be assessed as not changing the susceptibility to cyber attack, all other factors being equal.

By identifying the right set of dimensions, a disciplined assessment systematically guides an analyst through how well a future logistics concept provides versatility to the range of duress that logistics could face. Most assessments in this approach will not be quantitative. The nature of the uncertainties—e.g., the future threat environment and the variable maturity levels of different concepts and technologies—for the most part prevents assigning numerical values to a concept's contribution to versatility. As we will see in Chapter 4, where the framework is used in a case study, the product of each dimension generally will be a narrative. And given that the dimensions are not commensurate, we do not expect that the assessments will sum to a single estimate of increasing or decreasing versatility. Instead, we provide an assessment for each individual dimension.

The point of the framework is to enable analysts to systematically review for decisionmakers any proposed future logistics concept, identifying where versatility is expected to increase, where it is expected to decrease, and why. The goal is not to predict the future, but to show how any given concept would fare across the variables that can describe a potential future. It is to run the future logistics concept through an analytical gauntlet to reveal its operational benefits and weaknesses.

At the core of a framework for assessing the versatility of future logistics concepts lies the question of what the challenges are that the concepts might confront. This chapter explores that question in detail. Here, we expand the dimensions of the changing geopolitical setting, the threat environment, and evolving Blue CONEMPs into more-detailed factors.

The list of challenges expressed as dimensional factors is not meant to be predictive. These challenges should not be interpreted as likely scenarios that the U.S. Air Force will face. They have been developed to illustrate the degree to which any future logistics concept provides or does not provide versatility against all relevant environmental factors. Whether these factors are important or whether they justify the expenditure of resources to address is left to decisionmakers.

The list outlined in this chapter might need to be modified as experience accrues. Novel future logistics concepts might reveal a new aspect of versatility that has not yet been considered. Novel concepts could address a new aspect of versatility or expose the U.S. Air Force to a new aspect. In such cases, the list should be expanded to include new factors to cover these new elements.

Changing Geopolitical Setting

In a historically rare circumstance, over the past three decades, U.S. forces have been able to forward deploy, mass, and stage forces on friendly soil. They have been able to execute military operations at the time of their choosing, virtually unchallenged by the adversary. Although it is not clear exactly what future conflicts will look like, this particular era of sanctuary appears to have come to an end.[25] Formally documenting this change in the geopolitical setting, the NDS emphasizes the eroding U.S. military balance vis-à-vis Russia and China, among other adversaries, as well as the end of uncontested, all-domain superiority.[26] The NDS also formally documented the shift of U.S. defense planning away from terrorism and counterinsurgency campaigns toward competition and potential conflict with China and Russia.[27]

With this guidance in mind, and acknowledging that the actual future security environment may not follow this exact trajectory, we discuss the variety of subdimensions within the geopolitical setting that should be considered in an analysis of a future logistics concept. Specifically, this section addresses the implications of the intensity of conflict, the duration of high-intensity conflict, the potential for escalation, the geographic extent of conflict, dependency on allies and partners, and climate and geography.

Conflict Intensity

One of the most determinative factors with respect to the demand on logistics is the size of the deployed force and the operations tempo—the intensity of a potential conflict. The wars in

---

[25] Raphael S. Cohen, Nathan Chandler, Shira Efron, Bryan Frederick, Eugeniu Han, Kurt Klein, Forrest E. Morgan, Ashley L. Rhoades, Howard J. Shatz, and Yuliya Shokh, *The Future of Warfare in 2030: Project Overview and Conclusions*, Santa Monica, Calif.: RAND Corporation, RR-2849/1-AF, 2020.

[26] DoD, 2018, p. 3.

[27] DoD, 2018. A congressionally mandated commission concurred with the NDS assessment of the security environment. See Eric Edelman, Gary Roughead, Christine Fox, Kathleen Hicks, Jack Keane, Andrew Krepinevich, Jon Kyl, Thomas Mahnken, Michael McCord, Michael Morell, Anne Patterson, and Roger Zakheim, *Providing for the Common Defense: The Assessment and Recommendations of the National Defense Strategy Commission*, Washington, D.C.: U.S. Institute of Peace, 2018.

Iraq and Afghanistan capture the logistics requirements of a medium- to low-intensity conflict. However, a high-end fight is plausible in a future warfighting scenario, whether against a near peer or a less-capable adversary.[28] Military doctrine of the United States, China, and Russia all emphasize the necessity of high-intensity operations to seize the initiative, confuse the enemy, and establish control of the battlespace such that follow-on forces can surge into the theater.[29] The exact numbers of aircraft, munitions, personnel, etc., required for a high-intensity fight against a near peer (or even a less-capable adversary) are difficult to predict and will depend on the specifics of the particular conflict. Yet, the degree to which it supports high-intensity conflict should be central to the analysis of a particular materiel or nonmateriel solution for future logistics.

As the number of sorties that need to be flown and the number of targets that need to be destroyed increase, more personnel will need to be deployed and more food, fuel, munitions, and repairs will be required.[30] The intensity of a conflict thus stresses the ability to bring mass and firepower to bear. Key questions for versatility are (1) In a high-intensity environment—for example, in the event of heavy fire from long-range, conventional precision-guided munitions—does the future logistics concept enable the needed tempo of supply and resupply? and (2) Does it increase demand on scarce resources (e.g., maintenance, fuel, training) in a way that might hinder support of a high-intensity conflict?

---

[28] See for example, Scott Boston, Michael Johnson, Nathan Beauchamp-Mustafaga, and Yvonne K. Crane, *Assessing the Conventional Force Imbalance in Europe: Implications for Countering Russian Local Superiority*, Santa Monica, Calif.: RAND Corporation, RR-2402, 2018; Cohen et al., 2020; Gian Gentile, Yvonne K. Crane, Dan Madden, Timothy M. Bonds, Bruce W. Bennett, Michael J. Mazarr, and Andrew Scobell, *Four Problems on the Korean Peninsula: North Korea's Expanding Nuclear Capabilities Drive a Complex Set of Problems*, Santa Monica, Calif.: RAND Corporation, TL-271-A, 2019; David A. Shlapak and Michael W. Johnson, *Reinforcing Deterrence on NATO's Eastern Flank: Wargaming the Defense of the Baltics*, Santa Monica, Calif.: RAND Corporation, RR-1253-A, 2016; and David A. Shlapak, David T. Orletsky, Toy I. Reid, Murray Scot Tanner, and Barry Wilson, *A Question of Balance: Political Context and Military Aspects of the China-Taiwan Dispute*, Santa Monica, Calif.: RAND Corporation, MG-888-SRF, 2009.

[29] Chairman of the Joint Chiefs of Staff, Joint Publication 3-0, *Joint Operations*, Washington, D.C., August 11, 2011; Jeffrey Engstrom, *Systems Confrontation and System Destruction Warfare: How the Chinese People's Liberation Army Seeks to Wage Modern Warfare*, Santa Monica, Calif.: RAND Corporation, RR-1708-OSD, 2018, p. 15; Dave Johnson, *Russia's Conventional Precision Strike Capabilities, Regional Crises, and Nuclear Thresholds*, Livermore, Calif.: Lawrence Livermore National Laboratory Center for Global Security Research, Livermore Papers on Global Security No. 3, February 2018; Joshua Rovner, "Two Kinds of Catastrophe: Nuclear Escalation and Protracted War in Asia," *Journal of Strategic Studies*, Vol. 40, No. 5, 2017, pp. 697–698; Shou Xiaosong, ed., *The Science of Military Strategy*, Beijing: Military Science Press, 2013.

[30] For example, in the first 44 days of the air campaign in Operation Desert Storm, the Air Force flew 37,000 interdiction, air support, and counter air sorties. These sorties were supported by more than 11,000 refueling sorties. More than 60,000 tons of bombs were dropped. Moreover, the Air Force lost only 14 aircraft in the course of Desert Storm—a loss of 0.4 aircraft per 1,000 sorties. See Anthony H. Cordesman and Abraham R. Wagner, *The Lessons of Modern War*, Volume IV, *The Gulf War*, Boulder, Colo.: Westview Press, 1996, pp. 378, 380, 402–403, 492.

Conflict Duration

The ability of a future logistics concept to support a potentially long-duration conflict also must be considered. From 1946 to 2005, interstate conflicts lasted, on average, more than 200 days.[31] Certainly, the stakes would have to be very high for a conflict to break out, let alone for a prolonged high-intensity conflict. Yet a variety of factors suggest that a future conflict between the United States and a potential adversary could be lengthy and high-intensity. Domestic constraints, including domestic politics, nationalism, and the unpredictable nature of "stakes" and "interests" may make it difficult for leaders to back down or terminate the conflict.[32] In addition, the vast strategic depth available to the United States and several potential adversaries uniquely allows sanctuary that can be leveraged to continue a fight.[33] A potential future conflict might break out over indivisible issues—for example, ownership of or governing authority over a particular piece of land—which may prolong a conflict by provoking national pride or survival.[34] All of these factors suggest that numerous pathways to protracted, high-intensity war exist.[35] We do not argue that future conflicts will necessarily be protracted, but we do argue that the degree to which a future logistics concept can support protracted operations is a factor to consider. This possibility raises questions for a future concept. Specifically, what are the sustainment demands of the future logistics concept? Can the concept be maintained in the field for extended periods?

Escalation

Some future logistics concepts might be considered provocative by a potential adversary. We gave one example in Chapter 1: Prepositioning military materiel near a country could be seen as preparatory for war and might provoke unwanted escalation by that potential adversary. Other actions could include building or enhancing the capabilities of runways and other facilities that

---

[31] Joakim Kreutz, "How and When Armed Conflicts End: Introducing the UCDP Conflict Termination Dataset," *Journal of Peace Research*, Vol. 47, No. 2, 2010, p. 246. The intensity of these conflicts is not specified.

[32] Sarah E. Croco, *Peace at What Price? Leader Culpability and the Domestic Politics of War Termination*, Cambridge, Mass.: Cambridge University Press, 2015; George W. Downs and David M. Rocke, "Conflict, Agency, and Gambling for Resurrection: The Principal-Agent Problem Goes to War," *American Journal of Political Science*, Vol. 38, No. 2, May 1994; H. E. Goemans, *War and Punishment: The Causes of War Termination and the First World War*, Princeton, N.J.: Princeton University Press, 2000; H. E. Goemans and Mark Fey, "Risky but Rational: War as an Institutionally Induced Gamble," *Journal of Politics*, Vol. 71, No. 1, January 2009; Daniel Goure, "Moscow's Visions of Future War: So Many Conflict Scenarios So Little Time, Money and Forces," in Roger N. McDermott, ed., *The Transformation of Russia's Armed Forces: Twenty Lost Years*, New York: Routledge, 2015, pp. 124–125; Elizabeth A. Stanley, *Paths to Peace: Domestic Coalition Shifts, War Termination, and the Korean War*, Stanford, Calif.: Stanford University Press, 2009.

[33] Robert S. Ross, "The Geography of the Peace: East Asia in the Twenty-First Century," *International Security*, Vol. 23, No. 4, Spring 1999.

[34] Douglas M. Gibler, and Steven V. Miller, "Quick Victories? Territory, Democracies, and Their Disputes," *Journal of Conflict Resolution*, Vol. 57, No. 2, April 2013; Ron E. Hassner, "The Path to Intractability: Time and the Entrenchment of Territorial Disputes," *International Security*, Vol. 31, No. 3, Winter 2006/2007.

[35] Rovner, 2017, pp. 707–712.

the United States might use in a conflict. These escalatory considerations lead to a key question: Could the future logistics concept be perceived by an adversary or ally as provocative or escalatory?

A future logistics concept could prove escalatory during a conflict as well. Will a future logistics concept place the United States in a position such that it, or its allies, escalate because an adversary attacks it? The idea of forward deploying small nuclear reactors in theater to provide an energy source for U.S. forces is an illustrative example. A nuclear reactor would provide a military target for an adversary. Also, given the presence of nuclear materials, such a strike could be perceived as highly escalatory and could be interpreted as "crossing the nuclear threshold," which would almost certainly raise questions about whether the United States or the host nation should retaliate in kind.

A deployable nuclear reactor also could present an appealing target if an adversary did, in fact, want to signal a willingness to escalate the conflict. Alternatively, an adversary may assess that such an attack would be rife with the potential for escalation, and thus might avoid attempting to strike such a reactor. However, a forward-deployed reactor might nonetheless be targeted, either accidentally or by a commander in the field acting without permission, and the result could be an accidental or inadvertent escalation of the conflict. Analysis of future concepts thus should include the question, Does it increase or decrease the likelihood of intentional, unintentional, or accidental escalation of a conflict? Or, more broadly, does the future logistics concept have the potential to alter the strategic landscape with potential adversaries, and, if so, how?

Geographic Extent

The above examples represent cases of *vertical escalation*—i.e., escalation in the intensity of a conflict. Another escalation pathway that also must be considered is *horizontal escalation*— i.e., an adversary broadening the geographic scope of the conflict or an opportunistic third party taking advantage of the situation. At present, neither Russia nor China is known to be developing plans and strategies for opening up a second front in a potential conflict.[36] At the same time, China's pursuit of a blue-water navy and the expansive Belt and Road Initiative are creating the potential for it to obtain the ability to project forces and open up a second front.[37] Opportunistic aggression is also plausible. For example, a series of Russian military exercises suggest that it is

---

[36] Michael S. Chase, Jeffrey Engstrom, Tai Ming Cheung, Kristen A. Gunness, Scott Warren Harold, Susan Puska, and Samuel K. Berkowitz, *China's Incomplete Military Transformation: Assessing the Weaknesses of the People's Liberation Army (PLA)*, Santa Monica, Calif.: RAND Corporation, RR-893-USCC, 2015, p. 92; Fredrik Westerlund and Susanne Oxenstierna, eds., with Gudrun Persson, Jonas Kjellén, Johan Norberg, Jakob Hedenskog, Tomas Malmlöf, Martin Goliath, Johan Engvall, and Nils Dahlqvist, *Russian Military Capability in a Ten-Year Perspective—2019*, Swedish Defense Research Agency (FOI), FOI-R--4758--SE, December 2019, pp. 140–141.

[37] Evan S. Medeiros, "The Changing Fundamentals of US-China Relations," *Washington Quarterly*, Vol. 42, No. 3, 2019, p. 99.

preparing to overrun Georgia should a conflict between the United States and Iran break out.[38] Chinese political leaders and military doctrine also emphasize the importance of being willing to "discern opportunities to consolidate gains."[39] A country like North Korea could take advantage of heavy U.S. conflict with a near peer to seize the initiative on the Korean Peninsula.

Although a future conflict may stay contained within a single region, U.S. extended deterrence and security commitments mean that logistics planners must always be conscious of the requirement for projecting forces vast distances from the continental United States across more than one theater. These factors suggest that, when assessing a future concept, one must consider the following questions: To what degree does it improve capabilities to support operations in multiple theaters, including the continental United States? Will it increase or decrease the burden placed on U.S. force-projection capacity? Is it resilient to changes in the operational environment requiring a shift of resources to the front?

Dependency on Allies and Partners

The NDS emphasizes the centrality of allies and partners to the ability of U.S. forces to deter conflict and prevail in the event that deterrence fails.[40] When allies and partners host U.S. forces, it reduces the tyranny of time and distance faced by U.S. forces attempting to get to the fight.[41] Their approval and participation also strengthens the legitimacy and moral authority of the United States when it uses force.[42] With reliance, though, comes vulnerability—to both adversary actions and allied decisions.

Russia and China seek to cleave the United States from its allies and partners using economic and military coercion.[43] Providing U.S. forces with access is ultimately a political decision. An adversary thus might attempt to influence this decision by threatening to launch a long-range strike on an ally's territory. Previous commitments to host U.S. forces may not be upheld when that state is suffering costs for abiding by those commitments. Blue CONEMPs that call for

---

[38] Daniel Goure, "Moscow's Visions of Future War: So Many Conflict Scenarios So Little Time, Money and Forces," *Journal of Slavic Military Studies*, Vol. 27, No. 1, 2014, pp. 93–94.

[39] Timothy R. Heath, Kristen Gunness, and Cortez A. Cooper, *The PLA and China's Rejuvenation: National Security and Military Strategies, Deterrence Concepts, and Combat Capabilities*, Santa Monica, Calif.: RAND Corporation, RR-1402-OSD, 2016, pp. 19–20, 27.

[40] DoD, 2018, pp. 8–9.

[41] Thomas G. Mahnken, Grace B. Kim, and Adam Lemon, *Piercing the Fog of Peace: Developing Innovative Operational Concepts for a New Era*, Washington, D.C.: Center for Strategic and Budgetary Assessments, 2019, pp. 13–16.

[42] Patricia A. Weitsman, *Waging War: Alliances, Coalitions, and Institutions of Interstate Violence*, Palo Alto, Calif.: Stanford University Press, 2013.

[43] Ben Connable, Stephanie Young, Stephanie Pezard, Andrew Radin, Raphael S. Cohen, Katya Migacheva, and James Sladden, *Russia's Hostile Measures: Combating Russian Gray Zone Aggression Against NATO in the Contact, Blunt, and Surge Layers of Competition*, Santa Monica, Calif.: RAND Corporation, RR-2539-A, 2020; Mahnken, Kim, and Lemon, 2019, pp. 42–43.

more, smaller, and distributed force packages ostensibly mean that U.S. forces would require more access in a greater number of countries. Will the United States be able to obtain and sustain that access in the midst of adversary attacks on the host country?

The case of small deployable nuclear reactors is again a good example. For a future logistics concept of using these reactors for power in deployed locations, many nations might balk at the presence of any nuclear reactor on their soil or place severe restrictions on their use. Dependence on such a future logistics concept could diminish ally and partner nation support of U.S. forces.

A more mundane consideration, but an important one nonetheless, is that absent a preexisting agreement to host U.S. forces, the process to negotiate the necessary country clearances can be lengthy. In a situation in which timeliness could decide the intensity, duration, or outcome of a conflict, swiftly obtaining an ally's approval to provide access could be determinative. Key questions for versatility include the following: How might allies and partners respond to the concept? How might it change the willingness of the government (or the domestic populace) of an ally or partner nation to host U.S. forces? How reliant is the concept on being able to forward deploy troops, materiel, support, etc.? How vulnerable is the country to efforts by an adversary to coerce the state into withdrawing support from the United States?

Climate and Geography

The characteristics of a region heavily influence logistics requirements. Climate, weather patterns, and other environmental factors can affect the operational tempo, maintenance needs, and repair requirements. Weather events can place logistics operations at risk. Geography determines the availability of a local energy source and affects the ease of deployment and resupply. Geography also may be uniquely inhospitable to U.S. forces and weapons systems. Operations in the Arctic are a good example.[44] Both Russia and China highly value maintaining access to the Arctic theater (mostly for sea-based forces), viewing the theater as crucial for retaining confidence in the ability to escalate to the nuclear level.[45] Thus, it is not implausible that the United States would need to fight or patrol there. These factors suggest the questions: How resilient is the future concept to different climates? How resilient is it to the local availability of energy sources?

---

[44] Department of the Air Force, *Arctic Strategy: Ensuring a Stable Arctic Through Vigilance, Power Projection, Cooperation, and Preparation*, Washington, D.C., July 2020.

[45] Stephen J. Blank, "Enter Asia: The Arctic Heats Up," *World Affairs*, Vol. 176, No. 6, March–April 2014; Heather A. Conley and Caroline Rohloff, *The New Ice Curtain: Russia's Strategic Reach in the Arctic*, Washington, D.C.: Center for Strategic and International Studies, August 27, 2015; Lincoln E. Flake, "Russia's Security Intentions in a Melting Arctic," *Military and Strategic Affairs*, Vol. 6, No. 1, March 2014; Office of the Secretary of Defense, *Annual Report to Congress: Military and Security Developments Involving the People's Republic of China 2019*, Washington, D.C.: U.S. Department of Defense, May 2, 2019, pp. v, 114; O. S. Tanenya and V. N. Uryupin, "Certain Aspects of Employing the Airborne Forces in Russia's Arctic Zone," *Military Thought*, Vol. 8, No. 1, 2019; David C. Wright, *A Dragon Eyes the Top of the World: Arctic Policy Debate and Discussion in China*, Newport, R.I.: U.S. Naval War College, China Maritime Studies Institute, August 2011; Kevin Xie, "Some BRICS in the Arctic: Developing Powers Look North," *Harvard International Review*, Vol. 36, No. 3, Spring 2015.

Threat Environment

Combat service support operations will need to withstand a potentially wide range of threats. Exactly how these threats will evolve is uncertain. In this section, we describe dimensional factors on which to assess how versatile future logistics concepts are to these uncertainties.

It is important to bear in mind that, although threats can be categorized by the broad types in this section, the threats described in the NDS extend to all parts of the logistics enterprise. An adversary can attack U.S. Air Force logistics directly. It can also attack the deeper DoD logistics enterprise, including the depots and the Defense Logistics Agency. Attacks very well could extend to the commercial sector, the U.S. industrial base, and foreign partners. Furthermore, attacks could target supporting infrastructure, like power and communications, whether they are undersea, on the Earth's surface, or in space.[46]

Conventional Kinetic Threats

Conventional kinetic threats, such as guided missiles and bombs targeting aircraft, runways, dormitories, and other critical infrastructure, often are foremost on the mind. This threat category raises the following question: Does the future logistics concept change the ability to absorb a conventional kinetic attack and recover from it? Dispersing equipment geographically, the hardening of structures, and the use of camouflage, concealment, and deception all would increase versatility to the extent that they would reduce the susceptibility to kinetic attacks. Concentrating assets into fewer targets and placing assets within known striking range of the adversary would decrease versatility.

Nonkinetic Threats

Nonkinetic threats come in various forms. For logistics, chief among these threats are cyber threats, electronic warfare (EW), and information operations. An adversary can directly disrupt logistics processes via cyber operations in many ways. Access to computer systems can be denied. Communications can be interrupted. Data can be corrupted. Available data rates can be reduced (by cyber attack or jamming). And data can be permanently deleted. Cyber attacks can spread beyond the intent of the attacker and can have devastating effects on systems that were not even targeted.[47] The following are key questions for versatility: Is a future logistics concept increasing or decreasing the exposure to cyber attack? Is it creating a smaller target set for the adversary, or even a single point of failure? How exposed is the concept to EW? How exposed is the concept to information operations?

---

[46] Defense Intelligence Agency, *Challenges to Security in Space*, Washington, D.C., 2019.

[47] Andy Greenberg, "The Code That Crashed the World: The Untold Story of NotPetya, the Most Devastating Cyberattack in History," *Wired*, September 2018.

Beyond these direct threats, combat service support needs to be mindful of emissions and administrative signatures. High-end adversaries have sophisticated EW capabilities. Systems emitting electronic signatures can expose the location and nature of activities, increasing their likelihood of being attacked. In addition, logistics data traversing nonsecure networks might be monitored or manipulated by an adversary. These administrative signatures can reveal operations and, in the age of machine learning, potentially could reveal future operations to an adversary. Does a future logistics concept increase or decrease emissions signatures? What are the potential impacts of those signatures?

Special Operations Forces Threats

High-end adversaries generally have very capable special operations forces. Latent special operations forces can lurk near a location from which the U.S. Air Force might operate and strike at a critical juncture. The range of potential attacks is quite broad and could include sabotage against every part of logistics infrastructure and personnel. The key question for the versatility of a future logistics concept is: Does the concept increase or decrease the exposure of logistics equipment, infrastructure, processes, and personnel to special operations attacks? For example, concepts that would have U.S. Air Force logistics operate with fewer personnel and resources, and therefore with less equipped security forces, would decrease the ability to meet special operations threats, which would negatively affect versatility. Concepts that would situate logistics within U.S. Air Force organic organizations would be more resilient, in general, than those that situate logistics in the commercial sector, which might not be willing to operate in a higher-threat environment, or which might be less protected in such an environment.

Insider Threats

Perhaps the most insidious threat is that of the insider.[48] The insider threat is "the potential for an individual who has or had authorized access to an organization's assets to use that access, either maliciously or unintentionally, to act in a way that could negatively affect the organization."[49] The organizations of concern are all organizations that are part of the overall logistics enterprise that supports U.S. Air Force operations. The insider threat is not confined to U.S. Air Force personnel. It also extends to the commercial sector and partner nations. Does the concept place more or fewer personnel in critical logistics positions that the U.S. Air Force (or the U.S. government) cannot directly vet? What aspects of U.S. Air Force logistics does this exposure place at risk?

---

[48] See, for example, Matthew Bunn and Scott D. Sagan, eds., *Insider Threats*, Ithaca, N.Y.: Cornell University Press, 2016.

[49] Michael Theis, Randall Trzeciak, Daniel Costa, Andrew Moore, Sarah Miller, Tracy Cassidy, and William Claycomb, *Common Sense Guide to Mitigating Insider Threats*, 6th ed., Pittsburgh, Pa.: Carnegie Mellon University, Software Engineering Institute Technical Note CMU/SEI-2018-TR-010, 2019, p. ix.

Nuclear Threats

Since the end of the Cold War, nuclear threats have been a low priority for conventional forces. That state of affairs is changing. The 2018 *Nuclear Posture Review* calls for reinvigorating nuclear readiness in the services:

> . . . Combatant Commands and Service components will be organized and resourced . . . to integrate U.S. nuclear and non-nuclear forces to operate in the face of adversary nuclear threats and employment. The United States will coordinate integration activities with allies facing nuclear threats and examine opportunities for additional allied burden sharing of the nuclear deterrence mission.[50]

This direction means that part of being versatile is being able to operate in a nuclear environment. For logistics, that means being able to operate through radiological conditions from fallout to, potentially, the effects of electromagnetic pulse. Does the future logistics concept make it easier or harder to operate in nuclear-weapon–effects environments, particularly radioactive fallout and electromagnetic pulse?

Biological Threats

It is possible that some adversaries could employ biological weapons, or that operations might need to be done during the outbreak of a communicable disease. How will the future logistics concept function in the face of attack via biological warfare? Will it make operations in this environment easier or harder?

Chemical Threats

It is possible that some adversaries could employ chemical weapons during a conflict. How will the future logistics concept function in the face of attack via chemical warfare? Will it make operations in this environment easier or harder?

Evolving Blue Operational Concepts of Employment

The earlier sections, "Changing Geopolitical Setting" and "Threat Environment," address the variety of challenges and changes presented by Red and the environment in which Blue must operate. But Blue also will change, and future logistics concepts must be versatile to those changes. Of course, many of the changes that Blue makes will be in response to Red, so there will be some overlap in issues. But some changes that Blue pursues might arise from other factors, such as technological opportunity.

Logistics is a support function, enabling operations. Operations, both in the U.S. Air Force and in the joint environment, continuously evolve. Although it can be argued that the evolution

---

[50] Office of the Secretary of Defense, *Nuclear Posture Review*, Washington, D.C.: U.S. Department of Defense, 2018, p. viii.

of operations should take into account the ability of logistics to support those operations, because logistics is a supporting function, logistics evolves to meet operational needs more so than operations evolve to conform to logistical capabilities. An important factor in the versatility of a future logistics concept is the degree to which it not only supports current operations but also can support a wide variety of potential future operational concepts.

The point of this dimension of flexibility is to assess how well a future logistics concept supports future CONEMPs. It is beyond the scope of this research to identify or propose future CONEMPs or to evaluate the likelihood of their adoption. We cite two potential future CONEMPs here to illustrate the concept of evaluating how well a future logistics concept supports CONEMPs. Key questions are

- What are the leading CONEMPs that a future logistics concept might need to support?
- What new demands for logistics do these new CONEMPs present?
- How well can a future logistics concept meet these new demands?

The more that a future logistics concept can support a variety of future CONEMPs, especially without eroding the ability to support current CONEMPs, the more versatile the concept is.

Dynamic Force Employment

The NDS calls for dynamic force employment, which means that employment of forces should be proactive and scalable, theater postures should be flexible, and forces should be readily maneuverable.[51] Dynamic force employment is meant to provide military leaders with more options to be "strategically predictable, but operationally unpredictable."[52] Agile Combat Employment (ACE) is a U.S. Air Force concept for dynamic force employment.[53] What demands might this place on logistics?

A full analysis of ACE concepts and their implications for logistics is also beyond the scope of this project. ACE concepts are evolving, so we make only high-level generalizations. At the heart of these concepts lies maneuver. Maneuver places stressors on logistics and the command and control of logistics in the form of

- the need to operate out of a large number of bases
- the agility to quickly move from one location to another
- the capability of the future logistics concept to function adequately
  - in an impeded communications environment
  - in austere locations.[54]

---

[51] DoD, 2018, p. 7.

[52] DoD, 2018, p. 5.

[53] U.S. Air Force, *Agile Combat Employment for Force Providers, Annex A to Adaptive Operations in Contested Environments*, Washington, D.C., June 11, 2020, Not available to the general public.

[54] See Don Snyder, Kristin F. Lynch, Colby Peyton Steiner, John G. Drew, Myron Hura, Miriam E. Marlier, and Theo Milonopoulos, *Command and Control of U.S. Air Force Combat Support in a High-End Fight*, Santa Monica, Calif.: RAND Corporation, RR-A316-1, 2021.

The more effective the concept is at supporting the variety of challenges of dynamic force employment, the more versatile it is.

Attritable Aircraft

Dynamic force employment introduces new concepts for employing existing forces. Another concept involves introducing significant numbers of new weapon systems. One salient example is to introduce attritable aircraft into the U.S. Air Force structure. Attritable aircraft can be reusable aircraft, ground-launched cruise missiles on one-way missions, or both, depending on operational need. Concepts for attritable aircraft vary, from small aircraft using a runway for launch and recovery to those that do not require runways for launch or recovery. The latter, for example, might be launched from trailers and recovered by parachute.[55] Attritable aircraft complicate an adversary's operations. From a force-laydown perspective, attritable aircraft can operate out of numerous small, geographically dispersed locations without runways. These locations might be used for short periods of time, much like transporter erector launchers for missiles.

Again, a full analysis of the implications of attritable aircraft for logistics is beyond the scope of this research. However, the implications for logistics are profound. In this case, new demands are placed on logistics in the form of

- the capability to operate out of locations without runways
- the need to operate out of a large number of locations
- the need to be mobile.

The better the concept can meet these demands, the more versatile it is for supporting attritable aircraft operations.

Many of the subdimensions that arise in these two examples of future CONEMPs overlap or to some degree duplicate ones listed above in the sections "Changing Geopolitical Setting" and "Threat Environment." The purpose of this dimension is not necessarily to reveal new challenges, but to ensure that key challenges of any new CONEMPs are raised and assessed. The fact that a future logistics concept addresses, or does not address, challenges posed by future CONEMPs under serious consideration is itself a key indicator of versatility. Any challenges should be highlighted to decisionmakers, and any identified strengths for future CONEMPs can be useful justifications for a concept when defending it in DoD or to Congress.

---

[55] Thomas Hamilton and David Ochmanek, *Operating Low-Cost, Reusable Unmanned Aerial Vehicles in Contested Environments: Preliminary Evaluation of Operational Concepts*, Santa Monica, Calif.: RAND Corporation, RR-4407-AF, 2020. See also Steve Trimble, "U.S. Air Force Launches Fielding Plan for Skyborg Weapons," *Aviation Week & Space Technology*, July 13–26, 2020.

Summary

The previous sections elaborate more fully the scope of challenges that logistics might need to confront. It is not our goal to assign likelihoods to any of the circumstances outlined for the three dimensions. For convenience, we list the dimensions in Table 2.1.

Table 2.1. Summary of Operational Versatility Dimensions

Dimension	Dimensional Factor
Changing geopolitical setting	• Conflict intensity • Conflict duration • Potential for escalation • Geographic extent of conflict • Dependency on allies and partners • Climate and geography
Threat environment	• Conventional kinetic • Nonkinetic (cyber operations, jamming, emissions signatures, etc.) • Special operations forces • Insiders • Nuclear • Biological • Chemical
Evolving Blue CONEMPs changes	• What are the leading CONEMPs that a future logistics concept might need to support? • What new demands for logistics do these new CONEMPs present? • How well can a future logistics concept meet these new demands?

The dimensions of operational versatility are not readily assessed quantitatively. They are meant to be assessed on a relative scale, in which the relation is to the status quo that a future logistics concept is meant to replace. This concept is shown schematically in Figure 2.1. For many dimensions, a future logistics concept will not change the expected versatility relative to the status quo and can be depicted along the zero-change vertical axis. Some will be assessed as more versatile and will plot on the right side, with a qualitative estimate of the magnitude of that change. Others will plot to the left, proportional to how they are expected to degrade versatility.

24

Figure 2.1. Schematic Depiction of Dimensional Factors for Assessing Change to Versatility

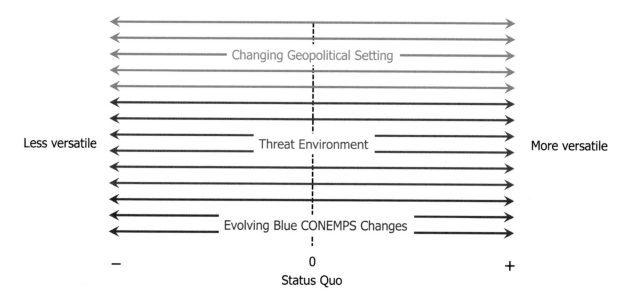

In Chapter 4, we demonstrate how to use these dimensional factors to assess versatility through selected case studies. But first, we turn to the issue of implementation challenges.

3. Assessing Implementation Challenges

> Technology does not dictate solutions. Rather, it provides a menu of options from which militaries choose. A service's culture, in turn, helps determine which options are more or less attractive.
>
> — Thomas G. Mahnken[56]

The versatility aspect of the Future Logistics Concept Assessment Framework, which we discussed in Chapter 2, is only one part of the overall evaluation needed to assess future logistics concepts. In addition to the operational benefits and downsides, the U.S. Air Force will need to consider the institutional disruptions necessary to implement a future logistics concept. That is, what would it take to transition from the status quo to the future logistics concept?

Not all future logistics concepts will simplify, consolidate, and reduce requirements. Higher operational versatility may well come at the expense of reduced efficiencies during peacetime. Moreover, the entire life cycle of a future logistics concept must be accounted for, including procurement, sustainment, and disposal costs. Some future concepts might induce a need to change organizational structure, culture, training, or other aspects of the U.S. Air Force institution, or perhaps those of entities outside the U.S. Air Force. These implementation challenges must be factored into decisions about a future logistics concept.

Formally exploring and documenting all the potential costs and other implementation challenges of a future logistics concept would be a considerable amount of work. What we advocate and outline in this chapter is a quick survey of the full range of challenges. The idea, much like the treatment of versatility in the last chapter, is to provide a quick guide to potential areas where there might be challenges in transitioning from the way logistics operates today to how it would need to work in a future including the future logistics concept. The approach will illustrate potential challenging areas and reveal how the future logistics concept is insufficiently defined to fully understand the challenges and, therefore, where risk might lie.

To organize the challenges, we follow the DOTMLPF-P framework.[57] The DOTMLPF-P framework was developed for just this purpose—to evaluate the impact that a joint concept is expected to have across all dimensions of DoD. Although we follow DOTMLPF-P as an organizing concept, we extend the discussion of many of the components of DOTMLPF-P beyond joint guidance. This extension is necessary to cover the range of potential challenges.

---

[56] Thomas G. Mahnken, *Technology and the American Way of War Since 1945*, New York: Columbia University Press, 2008, p. 11.

[57] Chairman of the Joint Chiefs of Staff Instruction 3010.02E, *Guidance for Developing and Implementing Joint Concepts*, Washington, D.C., August 17, 2016, directive current as of August 26, 2018.

The order of the topics that we discuss is the order used in the acronym DOTMLPF-P and does not reflect any particular priority.

Doctrine

Some future logistics concepts might be novel enough that doctrine will need to change to accommodate them. The degree of perturbation to doctrine can vary. Key questions include

- Does the future logistics concept contradict any existing doctrine?
- Do any allies or partner nations need to change doctrine?
- Does any doctrine need to change at the joint level?
- Does doctrine need to change only at the U.S. Air Force level?

Organization

It takes numerous organizations to implement a future logistics concept. Joint guidance focuses on changes to organizational structures, but the implementation challenges go beyond structural changes. For most future logistics concepts, many different organizations will need to play a role for the concept to be successfully developed and fielded. Their cooperation and agreement will be essential.

Implementing a future logistics concept will involve a large, diverse, complex, and often unwieldy enterprise. Many organizations within the Department of the Air Force will play significant roles. These organizations will need to agree to the concept and successfully execute their parts. Other organizations within DoD will also generally need to play a role, as well as organizations in other government agencies, and, perhaps, organizations in the commercial sector. Delays by any of these organizations, or failure to coordinate efforts across organizations, can imperil the implementation of a future logistics concept.

In general, the organizations that will play a role in a future logistics concept will change over time. Some might play a role early, during development or acquisition, and others might play a role later, during fielding and deployment. So, identifying the organizations and estimating the implementation challenges has a temporal dimension.

To assist in identifying organizations with key equities as a function of their situation and time, we offer a simple framework. The intent of this framework is not to build an exhaustive list of all of the entities within and outside the U.S. Air Force and DoD that could have equities for every potential concept. Nor is the purpose of this framework to cause the planner to build a staff summary sheet to solicit concurrence from all entities. Rather, the objective is to help planners identify the organizations that are expected to play a role in the concept's implementation. This transparency is essential in order to properly characterize the viability of any particular concept such that those responsible can make an informed decision.

Table 3.1 schematically depicts the approach for a notional example. In Table 3.1, the first axis of this framework, the columns, list the potential equities of the organizations within the

U.S. Air Force. From there, the analysis is broadened to consider the implementation equities of other organizations within DoD; other U.S. government organizations; and parties external to the government, such as private industry and partner nations. The second axis, the rows, frame the organizations that might have an implementation equity across the life cycle. Specifically, planners should consider the potential stakeholders at each of the following phases: research and development, acquisition, deployment and employment, sustainment, and disposition.

Table 3.1. Notional Logistics Implementation Equities

Phase	U.S. Air Force	Other DoD	Other U.S. Government	External
Research and development	AFRL	USD(R&E) DARPA		Commercial partners
Acquisition	AFMC • AFLCMC • AFIMSC	USD(A&S) OCIO	DOE NNSA	Commercial partners
Deployment and employment	ACC AMC AFGSC C-MAJCOMs AF/A10	Joint Staff/J4 COCOMs DLA	U.S. State Department DOE NNSA	Allies Partner nations
Sustainment	AFSC C-MAJCOMs	U.S. Army G4 DLA USTRANSCOM		Host nation
Disposition	AFMC • AFLCMC • AFIMSC	DLA		Host nation

NOTE: ACC = Air Combat Command; AF/A10 = Office of the Deputy Chief of Staff for Strategic Deterrence and Nuclear Integration; AFGSC = Air Force Global Strike Command; AFIMSC = Air Force Installation and Mission Support Center; AFLCMC = Air Force Life Cycle Management Center; AFMC = Air Force Materiel Command; AFRL = Air Force Research Laboratory; AFSC = Air Force Sustainment Center; AMC = Air Mobility Command; C-MAJCOM = Component Major Command; COCOM = Combatant Command; DARPA = Defense Advanced Research Projects Agency; DLA = Defense Logistics Agency; DOE NNSA = U.S. Department of Energy National Nuclear Security Administration; G4 = Deputy Chief of Staff for Army Logistics; J4 = Logistics; OCIO = Office of the Chief Information Officer; USD(A&S) = Office of the Under Secretary of Defense for Acquisition & Sustainment; USD(R&E) = Office of the Under Secretary of Defense for Research & Engineering; USTRANSCOM = U.S. Transportation Command.

For the organizational challenges of a future logistics concept, key questions include

- Which organizations are the primary stakeholders, and who has equities for each stage of the life cycle of the concept?
- What coordination is required across these organizations?
- Where are these organizations situated (e.g., in the U.S. Air Force, in DoD)?
- What are the risks of one of these organizations failing to perform its role?
- Will any of these organizations be expected to change its structure or culture to accommodate the concept?

Training

Often, a future logistics concept will require some degree of new training for personnel. Key questions include

- How will this future logistics concept affect training requirements?
- Will this future logistics concept affect existing training, affect cross-training, or require a whole new training course? If so, what is the training offset?

Materiel

Many future logistics concepts will involve materiel changes. Acquisition of materiel brings many implementation challenges. Some challenges arise directly from acquiring new materiel, but additional challenges derive from how legacy materiel will need to adapt to accommodate new materiel.

New Materiel

Any materiel acquisition takes time, money, and effort. The intent of this section is to flag potential challenges of which leaders need to be aware. We focus on indicators that the future logistics concept could consume considerable time and resources or is at risk of being derailed during development and acquisition. In this light, we first highlight common ways in which programs tend to stumble, then summarize simple ways to express the effort needed to acquire a new materiel solution.

Acquisition programs can falter for several reasons. They do not have to fail to the point of being canceled to be problematic. Programs that have significant issues with cost growth, schedule slippage, or performance deficiencies consume resources and time that could be used for other endeavors. These issues erode the credibility of the service. Analysis has highlighted two common sources of problems in acquisition programs: when programs are prematurely approved and when programs are poorly structured by an inadequate acquisition strategy.[58]

A key consideration for a future logistics concept is its technological maturity. *Technological maturity* refers to the readiness of a defined technology for application within a specific mission area. The standard measure is the technology readiness level (TRL).[59] We are not advocating a full technology readiness assessment for each future logistics concept, but we do recommend a

---

[58] Past RAND Corporation research identified two categories of common characteristics that lead to poor-performing programs: (1) premature approval of Milestone B and (2) suboptimal acquisition strategies and program structure that then impose high costs and cause schedule slippage. See Mark A. Lorell, Robert S. Leonard, and Abby Doll, *Extreme Cost Growth: Themes from Six U.S. Air Force Major Defense Acquisition Programs*, Santa Monica, Calif.: RAND Corporation, RR-630-AF, 2015; and Mark A. Lorell, Leslie Adrienne Payne, and Karishma R. Mehta, *Program Characteristics That Contribute to Cost Growth: A Comparison of Air Force Major Defense Acquisition Programs*, Santa Monica, Calif.: RAND Corporation, RR-1761-AF, 2017.

[59] Assistant Secretary of Defense for Research and Engineering, *Technology Readiness Assessment (TRA) Guidance*, Washington, D.C.: U.S. Department of Defense, April 2011.

general estimate of TRL to alert decisionmakers of the technological risk the concept might carry.

In addition to understanding the maturity of the future logistics concept, it is important to understand the acquisition strategy being proposed for the future logistics concept. For example, purchasing commercial off-the-shelf items can significantly reduce costs. However, this approach can lead to underestimations of the challenges associated with modifying and integrating commercially derived technologies for military use. Alternatively, the urgency of the operational need may decide the acquisition pathway, either as an Urgent Operational Need or a Joint Urgent Operational Need (JUON). The choice of contracts also can affect the program. Other Transaction Authorities (OTAs)—contracts that engage industry and academia for research and development, prototyping, and production—can be much faster and more flexible but they do not follow the same rules and regulations as other contract vehicles. As part of the assessment framework, it is important to identify the future logistics concept's acquisition strategy so decisionmakers can identify the challenges it might bring.

Key questions are

- Does the future logistics concept require further research and development? If so, what is the TRL?
- Must the future logistics concept pass through the full Joint Capabilities Integration and Development System and acquisition processes? Or
  - can it be purchased as a commercial off-the-shelf commodity?
  - is it a candidate for a JUON acquisition?
  - is it a candidate for OTA?
  - is it a candidate for Section 804 Authorities (Rapid Prototyping)?[60]
- Is the new materiel solution potentially life-limited because of technology obsolescence?
- What are the approximate life cycle costs to acquire, maintain, sustain, and dispose of the materiel?

Legacy Materiel

Virtually any new materiel acquisition will need to operate with existing legacy systems. Therefore, it might place different or higher demands on other systems. Communications needs provide an example. A new materiel acquisition might have specific communication needs and increase the data rate across existing communication channels. Another example might be munitions loaders. If attritable aircraft operate out of sites without a runway, will munitions loaders be needed that can operate off-road?

These considerations lead to additional questions, including

- To what extent does the new materiel depend on and interact with legacy materiel?
- What support does the new materiel need from existing systems?

---

[60] Public Law 114-92, *National Defense Authorization Act for Fiscal Year 2016*, November 25, 2015, Section 804.

- What communications capabilities does the new system require?

Leadership and Education

A future logistics concept could require changes in leadership and education. Most importantly, it might require changes to the command and control of logistics. Key questions include

- How will the future logistics concept affect reporting relationships and centralized and decentralized decisionmaking authority?
- Are any significant changes to command and control required, and what effects would that have on other logistics processes?

Personnel

The future logistics concept could require additional personnel. Key questions include

- Does the future logistics concept require new authorization levels, new positions or job titles, or new Air Force Specialty Codes (active duty, reserve component, civilians)?
- Does the future logistics concept require contractor support?
- Does the concept levy expectations for additional personnel from any partner nation?

Facilities

Some future logistics concepts might need changes to facilities or additional facilities. These changes could be required at the home station or at deployed locations, including any facilities for storage or in-transit support. Key questions include

- What military property, installations, and industrial facilities are needed for this future logistics concept (at home station and when deployed)?
- Will these facilities require military construction?
- Who will operate and maintain them?

Policy

Policies might need to be adjusted for a future logistics concept, but of more concern is whether the concept complies with all laws or regulations. Tactics, techniques, and procedures also might need to be changed. Key questions include

- What joint, interagency, coalition, U.S., or international law or regulations could affect or limit this future logistics concept?
- Will any changes be required to policies or tactics, techniques, or procedures within the Department of the Air Force?

Each future logistics concept likely will have multiple enabler implementation challenges. An organization may try to distinguish and understand each challenge by assigning a difficulty level, estimating a time frame of impact, or suggesting solutions. These solutions either could try to change the DOTMLPF-P or could change the future logistics challenge and would vary by case.

Alternatively, the future logistics concept may not be fully developed, and a planner may not be able to answer all of the questions posed here in detail. However, planners may be able to learn as much from admitting what they do not know about a concept as they can from reviewing what they do know.

For the U.S. Air Force to better assess a future logistics concept, it must consider the implementation challenges outlined in this chapter. In addition to the operational benefits and downsides (from Chapter 2), the Air Force will need to consider who has equities in the concept, what the acquisition considerations are (such as technology maturation and acquisition strategy), and what the challenges associated with DOTMLPF-P are. All of these are questions for consideration and will affect the feasibility and successful implementation of a future logistics concept.

To assist with the assessment of future logistics concepts, we present in the Appendix a checklist summarizing the assessment criteria discussed in this chapter and previously in Chapter 2. The checklist outlines a five-step process for evaluating a future logistics concept, which we demonstrate in a case study in the next chapter.

4. Case Study of the Assessment Processes

> ... [T]he first goal of a case study is to find out exactly what we have a case of.
> — Diane Vaughan[61]

In this chapter, we apply the Future Logistics Concept Assessment Framework described in Chapters 2 and 3 (and summarized in the Appendix) to an illustrative future logistics concept. We first discuss the purpose of the case study and then present the evaluation of the case study itself. It is important to note that we do not provide a recommendation for or against the future logistics concept. We merely provide the analysis, using a structured and transparent process. The output of the assessment is analysis, not advocacy for or against the concept.

Purpose of the Case Study

One purpose of this case study is to illustrate how the proposed framework can tease out the benefits and potential pitfalls of a future logistics concept. The application of the framework in this chapter shows that little technical knowledge is required for the assessments, and the process can be done rapidly. Yet the discipline that it imposes forces any advocate to confront a variety of issues.

A second purpose of the case study is to provide an exemplar of how an assessment might be done. We illustrate how to analyze a future logistics concept, no matter how well- or ill-defined. The evaluation itself, whether it is able to answer all of the questions or only a few, provides useful information to decisionmakers about the concept itself and about the potential impacts of adopting the concept. Applying the framework, we assess risks and benefits, but we do not advocate for or against the concept. The proposed approach is applicable across materiel and nonmateriel solutions. In this chapter, we evaluate a materiel solution as it exercises a wide variety of the features of the framework.

Example Case Study

The example future logistics concept that we examine is that of a deployable nuclear reactor, specifically a class called very Small Modular Reactors (vSMRs). This future logistics concept could help solve a challenge identified for supporting the NDS: a reliable power source for deployed locations. This future logistics concept is a real one. As of 2020, DoD is funding an

---

[61] Diane Vaughan, *The Challenger Launch Decision: Risky Technology, Culture, and Deviance at NASA*, Chicago, Ill.: University of Chicago Press, 1996, p. 56.

initiative called Project Pele to explore design prototypes.[62] Project Pele is in exploratory stages, and many details remain unspecified. This state will be common to many future logistics concepts. So, we will not be assessing Project Pele specifically, but we will discuss the general concept of using vSMRs for energy at deployed locations to support combat operations during wartime.

In the sections below, we walk through the Future Logistics Concept Assessment Framework—as described in Chapters 2 and 3 and summarized in the checklist in the Appendix—to evaluate this deployable future logistics concept.

Step 1: Define the Intended Operational Benefit of the Future Logistics Concept

The principal purpose of the vSMR future logistics concept is reliable power generation at deployed sites. The goals are (1) to reduce dependency on host-nation power; (2) to reduce dependency on supply lines for fossil fuels for diesel generators; and (3) to provide more power to support potential future systems, such as directed energy weapons, which might have high energy demands.

Reducing dependency on host-nation power grids is an important priority. Often, host nations do not have excess power to divert to U.S. forces. And, the power grid could be an adversary target, placing the reliability of power at risk. vSMRs would reduce these risks. Although the use of vSMRs would reduce U.S. military dependency on fossil fuels consumed by diesel generators, diesel fuel is generally only a small amount of the fossil fuel requirement at a forward location; the largest consumer of fossil fuel at most sites is the aircraft. The high power density of vSMRs makes them good candidates to meet the needs of power-hungry directed-energy weapons, should they be fielded.

Step 2: Define the Future Logistics Concept

The concept that we will evaluate is the provision of electrical power to deployed locations at wartime via vSMRs.[63] These vSMRs will produce 1–10 megawatts and be small enough to be transported by existing mobility aircraft and trucks (in standard shipping containers). A single site could use one or more vSMRs for power. We assume for the sake of argument that vSMRs would have a high power density relative to diesel generators, but that might depend on design details. Each vSMR will be able to operate within three days of arrival at the site and be installed below grade. For safety reasons, they will be passively cooled and be fueled by low-enriched uranium. They should be able to be shut down, sufficiently cooled, and removable from the deployed site in roughly one week. The vSMRs will not need refueling for years. Spent fuel will be removed in the United States, not at a deployed location.

---

[62] Bertuca, 2020.

[63] Defense Science Board, 2016; Trimble, 2019.

Several design ideas at varying stages of maturity exist for such a concept. We will assess this general concept, not any specific proposed design. As designs mature, the assessments can be refined.

Step 3: Evaluate Operational Versatility

Changing Geopolitical Setting

Conflict Intensity

The higher the intensity, the more forces the U.S. Air Force is likely to have in theater. The more forces in theater, the more electrical power will be needed. A high-intensity conflict likely will create a high demand for energy resources. A vSMR could provide a source of energy and thus increase the likelihood that U.S. forces will have the resources needed to meet this demand. Because they do not need a resupply of fuel, and assuming that they need no other substantial resupply, vSMRs are favorable for supporting a high operational tempo.

However, high-intensity conflicts also present a greater potential for an adversary to pursue a strategy that targets forward-deployed U.S. bases, attempting to deny the United States and its allies the ability to prevail in the conflict. Thus, high-intensity conflicts increase the likelihood that a vSMR would be targeted by an adversary (see the "Conventional Kinetic Threats" section below).

Conflict Duration

If the conflict lasts for less than ten years, the vSMRs should not require refueling. The longevity and assumed low resupply requirements are favorable for a long-duration conflict relative to diesel generators, but are roughly the same when compared with using the host-nation power grid (differences in threat exposure will be discussed below).

Escalation

As discussed in Chapter 2, a vSMR would provide a military target for an adversary, and a strike where there are nuclear materials could be interpreted as crossing the nuclear threshold, and thus could be considered escalatory. It also would present an appealing target if an adversary did, in fact, want to signal a willingness to escalate the conflict. Alternatively, an adversary may intentionally avoid an attack to avoid the potential for escalation. However, a forward-deployed reactor might nonetheless be targeted, either accidentally or by a commander in the field acting without permission, and the result could be an inadvertent escalation of the conflict.

Geographic Extent

The use of vSMRs in one theater would leave traditional power-generation capabilities, such as diesel generators, available to support operations in another theater should an opportunistic third-party adversary decide to take advantage of a conflict in another region. If the adversary broadens the geographic scope of the conflict and vSMRs need to be moved, vSMRs potentially could be moved within weeks, assuming that they are designed to be restarted after a field shutdown. The additional demand placed on airlift capability relative to using host-nation power grids could be impactful, depending on how many vSMRs were acquired, how many foreign basing agreements could be obtained, and how broadly dispersed U.S. forces were in the conflict. Using strategic airlift to move vSMRs would decrease the airlift capacity available for other requirements. In sum, given the power density of vSMRs, and assuming that they can be restarted after field shutdown, vSMRs provide a more rapid mobility to cover a wide geographic extent than the status quo.

Dependency on Allies and Partners

One benefit is that vSMRs would make the installation independent of host-nation power. But there are downsides.

Obtaining a partner nation's permission to deploy nuclear reactors with U.S. forces, even if it is a vSMR, almost certainly would complicate (if not preclude) the ability of U.S. forces to gain access to bases on foreign soil in Asia, Europe, and the Middle East. Overflight rights en route to the deployed site also would be an issue. In a 2019 public opinion poll in Japan, for example, respondents described nuclear energy as "dangerous" (69.0 percent) and "unsettling" (56.0 percent).[64] Australia has never allowed a nuclear power plant within its borders. Obtaining permission to deploy a vSMR will require bilateral agreements that are putatively even more complicated than effectively combining a Visiting Forces Agreement (VFA) with an agreement to permit U.S. nuclear-powered warships entry into foreign ports. The United States already faces challenges maintaining existing VFAs.[65] In the Indo-Pacific area of responsibility, for example, the sheer number of agreements that would have to be forged is daunting, given the stigma and fear associated with radioactive materials. It is difficult to see how approval to forward deploy a vSMR would enable or expedite those agreements.[66] Even if the U.S. Air Force did not intend to deploy a vSMR, the very existence of vSMRs in the U.S. Air Force could make a potential host nation nervous that the United States might bring in this capability without permission. Overall, deployable vSMRs would greatly complicate ally and partner nation cooperation relative to the status quo.

Weather and Geography

A vSMR could provide a source of energy for bases, both abroad and domestic, where traditional and alternative sources of energy are not easily accessible, and thus make it easier for

---

[64] Kei Yamada, "JAERO's Recent Public Opinion Survey on Nuclear Energy: Support Rises Somewhat for Restarting NPPs," Japan Atomic Industrial Forum, Inc., webpage, March 22, 2019.

[65] Caroline Baxter, "If U.S. Forces Have to Leave the Philippines, Then What?" *Foreign Policy Research Institute*, February 27, 2020.

[66] Defense Science Board, 2016, p. 38.

U.S. forces to operate in otherwise inhospitable climates and regions. However, it is unclear how Arctic or even subtropical environments would affect vSMR performance.

Threat Environment

Conventional Kinetic Threats

One vSMR on a base provides a target for the adversary, and there could be more than one at a base. A vSMR would require an increased security footprint, which would mean more people and an increased number of targets. Additionally, if a vSMR were targeted, the personnel sent to undertake the remediation efforts could potentially add even more targets. If a vSMR were to be directly struck, depending on radiation release, the base might have to be abandoned or the host nation might force extraction of U.S. forces. Resources, including mobility resources, might be consumed for remediation of the site and equipment.

Transport of a vSMR when it is removed from a location creates a target of opportunity if it is moved during hostilities. These consequences are unique to nuclear power and greatly reduce versatility relative to the status quo.

Nonkinetic Threats

Without details about the design of a vSMR, it is not possible to assess whether it will change the susceptibility to cyber attacks or EW compared with existing energy sources. If the system is bought as commercial off-the-shelf technology, DoD will have little leverage over incorporating cybersecurity into the design. Instead, DoD will rely on security controls on an existing design. If the system is acquired as a government program, cybersecurity requirements can be levied.

The consequences of a failure because of cyber attack could be higher for vSMRs than for the status quo because of the possibility of radioactive release. And even if a vSMR is designed to "fail safe," it may increase the *perception* of a vulnerability because of fears about a reactor meltdown being triggered by a cyber attack.

Assuming atypical security protocols to transport the vSMR to the base (relative to alternative sources of energy or materiel) and the likely increase in security personnel on the base, a vSMR would have a greater emission signature—that is, an adversary would have a greater chance of finding the base.

The deployment of a vSMR also would be a natural target of adversarial information warfare. An adversary might attempt to foment fear in the host nation's populace with respect to radioactive materials. It is not implausible that such a domestic backlash could lead to the termination of existing VFAs (see the earlier section on "Dependency on Allies and Partners").

Special Operations Forces Threats

A vSMR could create an opportunity for sabotage by adversary special operations forces. Relative to a power grid, vSMRs would be less exposed. But successful attacks on vSMRs could have higher consequences than attacks on power grids or diesel generators because of the potential for radioactive contamination.

Insider Threats

A vSMR increases the number of people on base, which increases the opportunity for insiders to infiltrate and cause harm. With an increased number of people, the number of systems that they could sabotage could increase as well. If the vSMR depends on contractor support, perhaps by reachback, the insider threat would be increased.

Nuclear Threats

If it is successfully targeted by an adversary and the containment vessel were breached, a vSMR could create the equivalent of a "dirty bomb." If an adversary were to intentionally or inadvertently strike the reactor, there is a potential for the perception that the nuclear threshold has been crossed (see the "Escalation" section above). Furthermore, if the U.S. Air Force had to fight in a nuclear weapons–effects environment, the approximate week that a vSMR takes to shut down and cool before being able to be moved means that the base population would not be able to move quickly to flush in advance of fallout from a nuclear plume. Another consideration is whether vSMRs would require hardening to early time electromagnetic pulse and the associated costs of that hardening. All things being equal, a vSMR would be less susceptible to late-time electromagnetic pulse (and extreme space weather) relative to a typical power grid.

Chemical Threats

Other than a potential situation in which personnel might need to evacuate a location in haste, leaving the vSMR behind, no change relative to the status quo is expected.

Biological Threats

Other than a potential situation in which personnel might need to evacuate a location in haste, leaving the vSMR behind, no change relative to the status quo is expected.

Evolving (Joint) Blue Concepts of Employment

In this section, we address three potential changes to Blue CONEMPs and the versatility of vSMRs to meet them: (1) increased use of maneuver, (2) the fielding of attritable aircraft, and (3) the fielding of directed energy weapons that are energy-intensive.

Maneuver

Maneuver places demands on speed of movement from one location to another, requires a light footprint, and potentially disperses operations over a larger number of locations than otherwise.

The vSMR concept requires a minimum of three days to set up and seven days to remove. Relative to the status quo, it is not agile for speedy movement from location to location. In addition, some site preparation is needed to situate the vSMRs below grade. These factors limit maneuver flexibility relative to the status quo.

The high power density of vSMRs lowers their footprint relative to that of diesel generators (for a given power level), but not relative to using a host nation's power grid.

Because of the probable reluctance of countries to permit vSMRs in their territories or the lengthy time to negotiate agreements to allow their use, this future logistics concept reduces flexibility to deploy to a diversity of locations across many nations (see the "Dependency on Allies and Partners" section above). With vSMRs, the U.S. Air Force is expected to have reduced ability to disperse forces relative to the status quo. Also, the more locations from which the U.S. Air Force operates, the more vSMRs will be needed. More vSMRs create more targets for the adversary and a higher risk of radioactive contamination (see the "Conventional Kinetic Threats" section above).

Attritable Aircraft

For future CONEMPs that use attritable aircraft launched from a trailer and recovered by parachute (if they are recovered), no runway is required. Freed of a runway, operations can be small and nimble. Such locations are likely to have small energy demands. The power density of vSMRs is not as useful in this case, and all of the considerations discussed above regarding maneuver render vSMRs cumbersome in these circumstances.

Directed Energy Weapons

vSMRs would increase the capabilities of U.S. forces to meet the increased energy demand created by novel weapons systems that are highly energy-intensive, such as directed-energy systems. They also could support other future systems that might demand high power, such as rail guns; water treatment and production; computing and data processing resources to support intelligence, surveillance, and reconnaissance; and additive manufacturing.

Step 4: Identify Implementation Challenges

Doctrine

Some additional U.S. Air Force and joint doctrine would need to be written, including doctrine about interacting with allies and partner nations.

Organization

Organizational Changes

A nuclear organization (e.g., flight, squadron) might be needed at the wing level. Other organizations might need to be added to support the vSMR above the wing level.

Organizational Equities

Several organizations would have equities in implementing a vSMR future logistics concept.

Focusing first on the equities within the U.S. Air Force, the A4 staffs at Headquarters U.S. Air Force, Air Combat Command, Air Mobility Command, and Global Strike Command all would have an interest in how a vSMR concept might be implemented. Tasked with making logistics-related policy, the staff of the U.S. Air Force Deputy Chief of Staff for Logistics, Engineering, and Force Protection will have equities in how vSMRs might affect the other functional areas that are tasked with providing combat service support. Air Combat Command's mission to organize, train, and equip U.S. Air Force forces will be affected by the introduction of special nuclear materials in the day-to-day operations during both peacetime and a high-end fight. The Air Mobility Command mission to provide rapid, global mobility and sustainment for U.S. armed forces would be affected by a decision to have mobility aircraft transport such a reactor to a forward location. For example, Air Mobility Command will need to develop any new flight procedures required to ensure the safe transport of a vessel containing radioactive materials, or establish a special unit for this purpose similar to the 62nd Airlift Wing's Prime Nuclear Airlift Force.

Beyond these straightforward points of interest within the U.S. Air Force, at a minimum, component major commands (C-MAJCOMs), Air Force Materiel Command (AFMC), Air Education and Training Command (AETC), and the U.S. Air Force Deputy Chief of Staff for Strategic Deterrence and Nuclear Integration (AF/A10) all could have equities throughout the vSMR life cycle. The C-MAJCOMs should be expected to have an interest at the acquisition, deployment, and sustainment stages. Specifically, the C-MAJCOMs would want to inform decisions regarding the quantity, transportability, security footprint, and hardness of the vSMRs in the event of kinetic attack. The C-MAJCOMs also would want to understand the required lead time before a vSMR could be forward deployed, which countries had approved overflight of an aircraft carrying such cargo, and which countries had approved hosting it.

Each of the six centers in AFMC also could have equities in such a concept, particularly with respect to their responsibilities for research and development, acquisition, and life cycle management. For example, AFMC and its centers would have an interest in decisions about how the fuel for a vSMR would be supplied, maintained, resupplied, and disposed of once it was spent. The AETC would be responsible for many facets at the deployment and sustainment stages, including providing the technical training on how to operate, safeguard, maintain, and transport the reactor. In peacetime—let alone in wartime—ensuring that the day-to-day workforce is trained to work around radiological material will be a nontrivial undertaking. Finally, Global Strike Command and the AF/A10 staff would have an interest in the implementation of such a concept, given that these offices might be called upon to develop a plan for how the United States might respond in the event that a forward deployed vSMR is targeted with a kinetic strike, whether purposefully or accidentally.

Looking outside the U.S. Air Force, several organizations within DoD would have equities at various points in the life cycle of implementing such a concept. The geographic combatant commands would have responsibility for the military use of vSMRs in their areas of responsibility. The Army could have interest in vSMRs and, given the history of the Army Nuclear Power Program, it could have additional insights from past experiences.[67] The Navy has expertise in nuclear power, and its assistance would be helpful for the success of the concept. U.S. Transportation Command and the Army's Military Surface Deployment and Distribution Command would want to understand the impacts on their responsibility for the land transport of trucks carrying such cargo. Similarly, the combat support agencies, in particular, the Defense Logistics Agency and Defense Threat Reduction Agency, would need to support this logistics concept. The Defense Logistics Agency could be tasked with the disposition of materials, and the Defense Threat Reduction Agency would have the mission to prepare for crisis response in the event of an accident or intentional strike on a vSMR.

Equities in implementing this future logistics concept also plausibly extend outside DoD. The U.S. Department of State and prospective host nations also would have an equity. The State Department serves as a key resource in negotiating and renegotiating partner nations' permission to deploy nuclear reactors with U.S. forces.[68] Host nations will have equities in terms of how the reactor will be secured, maintained, repatriated back to the United States, and how a potential radiological dispersal would be managed. If the State Department is unsuccessful in obtaining permission from potential host nations, vSMRs may be limited to in-garrison, domestic use.

Finally, the United States does not have a long-term storage solution for spent nuclear fuel. Although a significant amount of work has been completed in the review of building the Yucca Mountain Nuclear Waste Repository, final approvals have not been granted.[69] Given the absence of a long-term storage solution at the national level, at a minimum, the U.S. Nuclear Regulatory Commission will have an equity in the disposition of the spent nuclear fuel from these reactors.

Training

Training would be required for vSMR maintenance, support, and transportation. Other specialty training might be required, such as remediation of radioactive release.

Materiel

Concepts for vSMR are in development, at a TRL of approximately six, which means that a system/subsystem model or prototype has been demonstrated in a relevant environment.[70] A full-system prototype has yet to be demonstrated in a relevant environment, and some promises of

---

[67] Although the vSMR technologies are different from those employed by the Army from 1954 to 1976, there might be lessons from nontechnical aspects. See Lawrence H. Suid, *The Army's Nuclear Power Program: The Evolution of a Support Agency*, New York: Greenwood Press, 1990.

[68] Baxter, 2020.

[69] United States Nuclear Regulatory Commission, "High-Level Waste Disposal: NRC's Yucca Mountain Licensing Activities," webpage, updated March 12, 2020.

[70] Assistant Secretary of Defense for Research and Engineering, 2011.

safety have yet to be demonstrated.[71] The acquisition cost and life cycle sustainment costs for a deployable vSMR are uncertain and will depend on the acquisition strategy employed, which is also undetermined. Some additional support equipment might be required for vSMR maintenance, support, and transportation. Remediation of radioactive release also might require additional special equipment.

Leadership and Education

Education about vSMRs would need to be added to U.S. Air Force curriculum. Also, command and control of vSMRs would have to be integrated into the U.S. Air Force command and control system and processes.

Personnel

Personnel would have to be assigned to the vSMR mission. A new Air Force Specialty Code and career field might be needed to support the vSMR, or a shredout might be needed in the civil engineering career field. The acquisition strategy also might identify a need for contractor support for the vSMR capability.

Facilities

The United States does not currently possess a domestic uranium enrichment capability that can be leveraged to support U.S. military programs. If a treaty-compliant source of uranium fuel is not identified, a new production facility might be required. The vSMR will generate new support requirements, such as berms, pits, hardening, and camouflage, concealment, and deception. For safety, installations that use a vSMR probably will have to incorporate an emergency buffer zone.

Policy

Treaty obligations that preclude the United States from using foreign nuclear materials for defense purposes might have to be revised if domestic production of enriched uranium is not established or if other sources of fuel are identified. Laws and treaties to support bringing nuclear reactors onto foreign soil will have to be adopted. Will the vSMRs operate under the regulatory environment of the host nation, under those established by the U.S. Nuclear Regulatory Commission, or by special policies developed by DoD? Policies will need to be developed for the transportation and protection of nuclear components and fuels. Agreements with allies might be needed in advance for the use of vSMRs to reduce deployment time.

Additionally, new tactics, techniques, and procedures probably would need to be developed. These would cover both the operation of vSMRs and the contingency response to any radioactive contamination and remediation.

---

[71] Trimble, 2019.

Step 5: Present Findings to Decisionmakers

This analysis presents assessments of changes relative to the status quo, assessments of operational versatility, and the implementation challenges of moving from the status quo to the future logistics concept. A secondary result is the revelation of areas of uncertainty, or areas in which the future logistics concept might be better defined.

In the vSMR example, our analysis identified requirements that could be better specified to assess operational versatility and implementation challenges. Will vSMRs suffer any limitations in operations in extreme weather conditions? How will cybersecurity requirements be specified, and what are the associated risks? If a vSMR is to be moved from one deployed location to another, will vSMRs have the capability to restart after a field shutdown without being transported back to the United States for reconstitution? Finally, is our initial assumption correct? Will the vSMRs have significantly higher power density than diesel generators?

5. Afterword

Many new ideas for military operations will be proposed to help meet the national needs expressed in the NDS. Some of these ideas will be future concepts for logistics. We presented a framework for evaluating future logistics concepts in Chapters 2 and 3 of this report.

- Chapter 2 examined the *operational versatility* of novel concepts, by which we mean that logistics should be sufficiently flexible to support uncertain environments, but not break; that logistics should be able to absorb and recover from attacks; and that logistics be sufficiently adaptable to support evolving Blue CONEMPs. Furthermore, logistics should be sufficiently timely (i.e., agile) across all of these aspects.
- Chapter 3 examined the *implementation challenges* of any future logistics concept, by which we mean the effort and costs necessary to go from the status quo to the future logistics concept.

The Appendix shows how to use the framework on a real case study.

Applicability of the Framework

Although it is formulated for evaluating future logistics concepts, the framework that we presented is useful beyond logistics. No aspect of the framework restricts its application to logistics. Any concept for addressing NDS challenges can be assessed using this same framework. Indeed, logistics is part of fully integrated, joint (and combined) operations. Therefore, new concepts need to be evaluated—not in isolation, but as parts of joint (and combined) all-domain operations. This need lies at the heart of the section in Chapter 2 on the operational utility of a concept as part of other future Blue CONEMPs.

Target Users of the Framework

The use of the method we present is not restricted to any one setting or one part of the Department of the Air Force. Its simplicity and comprehensiveness gives it value for assessing any new concept in any organization. A C-MAJCOM might have a novel idea to meet a pressing warfighting challenge. Staff might wish to assess the concept before investing too many resources in it, or to be better prepared when presenting the idea outside the command. The Air Force Warfighting Integration Capability might be considering novel options for future warfighting. It, too, can benefit from the systematic assessment of this framework. A major command might be considering an alternative way to organize, train, and equip. What benefits accrue, and what unintended consequences might be lurking?

Because the method is simple, a small number of staff in any organization should be able to rapidly assess future concepts well enough to reveal the variety of operational benefits, downsides, and implementation challenges and decide whether to proceed with them.

Utility of the Framework

We intend that the analyst using the framework inform decisions, not make them. The framework reveals the operational versatility and implementation challenges of a future concept. It does not help make the value judgments of whether any benefits are worth the costs. Military and political decisionmakers need to make those judgments, informed by the assessments of the framework.

Armed with this analysis, the Department of the Air Force will be better equipped to

- refine future concepts at an early stage of development
- vet future concepts earlier to avoid investing too many resources in exploring concepts that are unworthy
- prevent poorly conceived concepts from reaching senior decisionmakers
- assess future concepts systematically across a comprehensive gauntlet of factors so that, if presented to the Office of the Secretary of Defense and Congress, the Department of the Air Force can convincingly defend them.

Appendix. Future Logistics Concept Assessment Framework Checklist

This five-step checklist is intended to provide a quick reference guide for anyone asked to evaluate future logistics concepts for the U.S. Air Force. It contains the overall criteria outlined in this report. More detail for each of the evaluation areas listed below can be found in the main report.

This checklist is not designed to be all-inclusive. Nor it is designed to replace existing acquisition processes. The intent is to provide a disciplined, systematic way to assess proposed future logistics concepts by running ideas through an analytical gauntlet to expose any weaknesses so that, if the U.S. Air Force chooses to pursue a future logistics concept, it is well prepared to defend it to DoD and Congress. The checklist contains the following steps:

1. Define the intended operational benefit of the future logistic concept. This is an aspirational view of what the new concept will achieve if it is successful.
2. Define the future logistics concept. For some concepts, there might be a lot of detail. For others, there may be many unknowns. Both are helpful to the evaluation of the concept.
3. Evaluate the operational versatility of the future logistics concept within three main areas (changing geopolitical setting, changes in the threat environment, and evolving joint Blue CONEMPs).

A. A changing geopolitical setting

- Conflict intensity

  - Does the future logistics concept enable the needed tempo of supply and resupply?
  - Does it increase demand on scarce resources (e.g., maintenance, fuel, training) in a way that might hinder support of a high-intensity conflict?

- Conflict duration

  - What are the sustainment demands of the future logistics concept?
  - Can the concept be maintained in the field for extended periods?

- Escalation

  - Could the future logistics concept be perceived by an adversary or ally as provocative or escalatory?
  - Does the future logistics concept have the potential to alter the strategic landscape with potential adversaries, and, if so, how?

- Geographic extent

  - To what degree does the concept improve capabilities to support operations in multiple theaters, including the continental United States?

- Dependency on allies and partners

- How might allies and partners respond to the concept?
- How might it change the willingness of the government (or of the domestic populace) of an ally or partner nation to host U.S. forces?
- How reliant is the concept on being able to forward deploy troops, materiel, support, etc.?
- How vulnerable is the country to efforts by an adversary to coerce the state into withdrawing support from the United States?

- Weather and geography

  - How resilient is the future concept to different climates?
  - How resilient is it to the local availability of energy sources?

B. Changes in the threat environment (extending to the commercial sector, the U.S. industrial base, and foreign partners; attacks could target supporting infrastructure, like power and communications, whether they are undersea, on the Earth's surface, or in space)

- Conventional kinetic threats

  - Does the future logistics concept change the ability to absorb and recover from a conventional kinetic attack?

- Nonkinetic threats

  - Is a future logistics concept increasing or decreasing the exposure to cyber attack? Is it creating a smaller target set for the adversary, or even a single point of failure?
  - How exposed is the concept to electronic warfare?
  - How exposed is the concept to information operations?
  - Does a future logistics concept increase or decrease emissions signatures? What are the potential impacts of those signatures?

- Special operations forces threats

  - Does the concept increase or decrease the exposure of logistics equipment, infrastructure, processes, and personnel to special operations attacks?

- Insider threats

  - Does the concept place more or fewer personnel in critical logistics positions that the U.S. Air Force (or the U.S. government) cannot directly vet?
  - What aspects of U.S. Air Force logistics does this exposure place at risk?

- Nuclear threats

  - Does the future logistics concept make it easier or harder to operate in nuclear-weapon–effects environments, particularly radioactive fallout and electromagnetic pulse?

- Biological threats

  - How will the future logistics concept function in the face of attack via biological warfare? Will it make operations in this environment easier or harder in any way?

- Chemical threats
  - How will the future logistics concept function in the face of attack via chemical warfare? Will it in any way make operations in this environment easier, or harder?

C. Evolving (joint) Blue CONEMPs

- What are the leading CONEMPs that a future logistics concept might need to support?
- What new demands for logistics do these new CONEMPs present?
- How well can a future logistics concept meet these new demands?

4. Identify implementation challenges associated with the future logistics concept in three main areas. The goal is to recognize—not solve—the challenges as early in the process as possible.

A. Doctrine

- Does the future logistics concept contradict any existing doctrine?
- Do any allies or partner nations need to change doctrine?
- Does any doctrine need to change at the joint level?
- Does doctrine need to change only at the U.S. Air Force level?

B. Organization

Which organizations are the primary stakeholders, and who has equities for each stage of the life cycle of the concept?

- What coordination is required across these organizations?
- Where are these organizations situated (e.g., in the U.S. Air Force, in DoD)?
- What are the risks of one of these organizations failing to perform their roles?
- Will any of these organizations be expected to change structure or culture to accommodate the concept?

C. Training

- How will this future logistics concept affect training requirements?
- Will this future logistics concept affect existing training, affect cross-training, or require a whole new training course? If so, what is the training offset?

D. Materiel

- Does the future logistics concept require further research and development? If so, what is the TRL?
- Must the future logistics concept pass through the full Joint Capabilities Integration and Development System and acquisition processes? Or,
  - can it be purchased as a commercial off-the-shelf commodity?
  - is it a candidate for a JUON acquisition?
  - is it a candidate for OTA?
  - is it a candidate for Section 804 Authorities (Rapid Prototyping)?

- Is the new materiel solution potentially life-limited because of technology obsolescence?
- What are the approximate life cycle costs to acquire, maintain, sustain, and dispose of the materiel?
- To what extent does the new materiel depend on and interact with legacy materiel?
- What support does the new materiel need from existing systems?
- What communications capabilities does the new system require?

E. Leadership and education

- How will the future logistics concept affect reporting relationships and centralized and decentralized decisionmaking authority?
- Are any significant changes to command and control required, and what effects would that have on other logistics processes?

F. Personnel

- Does the future logistics concept require new authorization levels, new positions or job titles, or new Air Force Specialty Codes (e.g., active duty, reserve component, civilians)?
- Does the future logistics concept require contractor support?
- Does the concept levy expectations for additional personnel from any partner nation?

G. Facilities

- What military property, installations, and industrial facilities are needed for this future logistics concept (at home station and when deployed)?
- Will these facilities require military construction?
- Who will operate and maintain them?

H. Policy

- What joint, interagency, coalition, U.S., or international law or regulations could affect or limit this future logistics concept?
- Will any changes be required to policies or tactics, techniques, or procedures within the Department of the Air Force?

5. Present the findings to decisionmakers. It is important to identify the portions of the framework that could not be completed as well as the portions that could be completed. Completed correctly, this framework will improve the situational awareness of senior leaders before they make decisions and defend them better afterward.

Abbreviations

ACE	Agile Combat Employment
AETC	Air Education and Training Command
AF/A10	U.S. Air Force Deputy Chief of Staff for Strategic Deterrence and Nuclear Integration
AFMC	Air Force Materiel Command
C-MAJCOM	component major command
CONEMP	concept of employment
COVID-19	coronavirus disease 2019
DoD	U.S. Department of Defense
DOTMLPF-P	doctrine, organization, training, materiel, leadership and education, personnel, facilities, and policy
EW	electronic warfare
JUON	Joint Urgent Operational Need
NDS	National Defense Strategy
OTA	Other Transaction Authority
TRL	technology readiness level
VFA	Visiting Forces Agreement
vSMR	very Small Modular Reactor

References

Assistant Secretary of Defense for Research and Engineering, *Technology Readiness Assessment (TRA) Guidance*, Washington, D.C.: U.S. Department of Defense, April 2011.

Baraniuk, Chris, "Blockchain: The Revolution That Hasn't Quite Happened," BBC News, February 11, 2020. As of February 19, 2020:
https://www.bbc.com/news/business-51281233

Baxter, Caroline, "If U.S. Forces Have to Leave the Philippines, Then What?" *Foreign Policy Research Institute*, February 27, 2020. As of April 8, 2020:
https://www.fpri.org/article/2020/02/if-u-s-forces-have-to-leave-the-philippines-then-what/

Bertuca, Tony, "DOD Awards Contracts for Prototype Mobile Nuclear Reactor," *Inside Defense*, March 9, 2020.

Blank, Stephen J., "Enter Asia: The Arctic Heats Up," *World Affairs*, Vol. 176, No. 6, March–April 2014, pp. 19–29.

Boston, Scott, Michael Johnson, Nathan Beauchamp-Mustafaga, and Yvonne K. Crane, *Assessing the Conventional Force Imbalance in Europe: Implications for Countering Russian Local Superiority*, Santa Monica, Calif.: RAND Corporation, RR-2402, 2018. As of June 1, 2021:
https://www.rand.org/pubs/research_reports/RR2402.html

Brown, Charles Q., Jr., *Accelerate Change or Lose*, Washington, D.C.: U.S. Air Force, August 2020.

Bunn, Matthew, and Scott D. Sagan, eds., *Insider Threats*, Ithaca, N.Y.: Cornell University Press, 2016.

Chairman of the Joint Chiefs of Staff, *DOD Dictionary of Military and Associated Terms*, Washington, D.C.: U.S. Department of Defense, January 2020.

Chairman of the Joint Chiefs of Staff, Joint Publication 3-0, *Joint Operations*, Washington, D.C., August 11, 2011.

Chairman of the Joint Chiefs of Staff Instruction 3010.02E, *Guidance for Developing and Implementing Joint Concepts*, Washington, D.C., August 17, 2016, directive current as of August 26, 2018.

Chairman of the Joint Chiefs of Staff Instruction 3030.01, *Implementing Joint Force Development and Design*, Washington, D.C., December 3, 2019.

Chase, Michael S., Jeffrey Engstrom, Tai Ming Cheung, Kristen A. Gunness, Scott Warren Harold, Susan Puska, and Samuel K. Berkowitz, *China's Incomplete Military Transformation: Assessing the Weaknesses of the People's Liberation Army (PLA)*, Santa Monica, Calif.: RAND Corporation, RR-893-USCC, 2015. As of June 2, 2021: https://www.rand.org/pubs/research_reports/RR893.html

von Clausewitz, Carl, *On War*, edited and translated by Michael Howard and Peter Paret, Princeton, N.J.: Princeton University Press, 1976.

Cohen, Raphael S., Nathan Chandler, Shira Efron, Bryan Frederick, Eugeniu Han, Kurt Klein, Forrest E. Morgan, Ashley L. Rhoades, Howard J. Shatz, and Yuliya Shokh, *The Future of Warfare in 2030: Project Overview and Conclusions*, Santa Monica, Calif.: RAND Corporation, RR-2849/1-AF, 2020. As of June 2, 2021: https://www.rand.org/pubs/research_reports/RR2849z1.html

Conley, Heather A., and Caroline Rohloff, *The New Ice Curtain: Russia's Strategic Reach in the Arctic*, Washington, D.C.: Center for Strategic and International Studies, August 27, 2015.

Connable, Ben, Stephanie Young, Stephanie Pezard, Andrew Radin, Raphael S. Cohen, Katya Migacheva, and James Sladden, *Russia's Hostile Measures: Combating Russian Gray Zone Aggression Against NATO in the Contact, Blunt, and Surge Layers of Competition*, Santa Monica, Calif.: RAND Corporation, RR-2539-A, 2020. As of June 2, 2021: https://www.rand.org/pubs/research_reports/RR2539.html

Cordesman, Anthony H., and Abraham R. Wagner, *The Lessons of Modern War,* Volume IV, *The Gulf War*, Boulder, Colo.: Westview Press, 1996.

Croco, Sarah E., *Peace at What Price? Leader Culpability and the Domestic Politics of War Termination*, Cambridge, Mass.: Cambridge University Press, 2015.

Defense Intelligence Agency, *Challenges to Security in Space*, Washington, D.C., 2019.

Defense Science Board, *Task Force on Energy Systems for Forward/Remote Operating Bases: Final Report*, Washington, D.C.: U.S. Department of Defense, Office of the Under Secretary of Defense for Acquisition, Technology, and Logistics, August 1, 2016.

Department of the Air Force, *Arctic Strategy: Ensuring a Stable Arctic Through Vigilance, Power Projection, Cooperation, and Preparation*, Washington, D.C., July 2020.

DoD—*See* U.S. Department of Defense.

Downs, George W., and David M. Rocke, "Conflict, Agency, and Gambling for Resurrection: The Principal-Agent Problem Goes to War," *American Journal of Political Science*, Vol. 38, No. 2, May 1994, pp. 362–380.

Edelman, Eric, Gary Roughead, Christine Fox, Kathleen Hicks, Jack Keane, Andrew Krepinevich, Jon Kyl, Thomas Mahnken, Michael McCord, Michael Morell, Anne Patterson, and Roger Zakheim, *Providing for the Common Defense: The Assessment and Recommendations of the National Defense Strategy Commission*, Washington, D.C.: U.S. Institute of Peace, 2018.

Engstrom, Jeffrey, *Systems Confrontation and System Destruction Warfare: How the Chinese People's Liberation Army Seeks to Wage Modern Warfare*, Santa Monica, Calif.: RAND Corporation, RR-1708-OSD, 2018. As of June 2, 2021:
https://www.rand.org/pubs/research_reports/RR1708.html

Erwin, Sandra, "Space Launch Vehicles Eyed by the Military to Move Supplies Around the World," *Space News*, August 2, 2018. As of February 19, 2020:
https://spacenews.com/space-launch-vehicles-eyed-by-the-military-to-move-supplies-around-the-world/

Flake, Lincoln E., "Russia's Security Intentions in a Melting Arctic," *Military and Strategic Affairs*, Vol. 6, No. 1, March 2014, pp. 99–116.

Gentile, Gian, Yvonne K. Crane, Dan Madden, Timothy M. Bonds, Bruce W. Bennett, Michael J. Mazarr, and Andrew Scobell, *Four Problems on the Korean Peninsula: North Korea's Expanding Nuclear Capabilities Drive a Complex Set of Problems*, Santa Monica, Calif.: RAND Corporation, TL-271-A, 2019. As of June 2, 2021:
https://www.rand.org/pubs/tools/TL271.html

Gibler, Douglas M., and Steven V. Miller, "Quick Victories? Territory, Democracies, and Their Disputes," *Journal of Conflict Resolution*, Vol. 57, No. 2, April 2013, pp. 258–284.

Goemans, H. E., *War and Punishment: The Causes of War Termination and the First World War*, Princeton, N.J.: Princeton University Press, 2000.

Goemans, H. E., and Mark Fey, "Risky but Rational: War as an Institutionally Induced Gamble," *Journal of Politics*, Vol. 71, No. 1, January 2009, pp. 35–54.

Goure, Daniel, "Moscow's Visions of Future War: So Many Conflict Scenarios So Little Time, Money and Forces," *Journal of Slavic Military Studies*, Vol. 27, No. 1, 2014, pp. 63–100.

Goure, Daniel, "Moscow's Visions of Future War: So Many Conflict Scenarios So Little Time, Money and Forces," in Roger N. McDermott, ed., *The Transformation of Russia's Armed Forces: Twenty Lost Years*, New York: Routledge, 2015.

Greenberg, Andy, "The Code That Crashed the World: The Untold Story of NotPetya, the Most Devastating Cyberattack in History," *Wired*, September 2018, pp. 52–63.

Hamilton, Thomas, and David Ochmanek, *Operating Low-Cost, Reusable Unmanned Aerial Vehicles in Contested Environments: Preliminary Evaluation of Operational Concepts*, Santa

Monica, Calif.: RAND Corporation, RR-4407-AF, 2020. As of June 2, 2021:
https://www.rand.org/pubs/research_reports/RR4407.html

Hassner, Ron E., "The Path to Intractability: Time and the Entrenchment of Territorial Disputes," *International Security*, Vol. 31, No. 3, Winter 2006/2007, pp. 107–138.

Heath, Timothy R., Kristen Gunness, and Cortez A. Cooper, *The PLA and China's Rejuvenation: National Security and Military Strategies, Deterrence Concepts, and Combat Capabilities*, Santa Monica, Calif.: RAND Corporation, RR-1402-OSD, 2016. As of June 2, 2021: https://www.rand.org/pubs/research_reports/RR1402.html

Horne, Alistair, *To Lose a Battle: France 1940*, Boston, Mass.: Little, Brown and Company, 1969.

Johnson, Dave, *Russia's Conventional Precision Strike Capabilities, Regional Crises, and Nuclear Thresholds*, Livermore, Calif.: Lawrence Livermore National Laboratory Center for Global Security Research, Livermore Papers on Global Security No. 3, February 2018.

Kent, Glenn A., with David Ochmanek, Michael Spirtas, and Bruce R. Pirnie, *Thinking About America's Defense: An Analytical Memoir*, Santa Monica, Calif.: RAND Corporation, OP-223-AF, 2008. As of June 2, 2021: https://www.rand.org/pubs/occasional_papers/OP223.html

Kreutz, Joakim, "How and When Armed Conflicts End: Introducing the UCDP Conflict Termination Dataset," *Journal of Peace Research*, Vol. 47, No. 2, 2010, pp. 243–250.

Lorell, Mark A., Robert S. Leonard, and Abby Doll, *Extreme Cost Growth: Themes from Six U.S. Air Force Major Defense Acquisition Programs*, Santa Monica, Calif.: RAND Corporation, RR-630-AF, 2015. As of July 23, 2020: https://www.rand.org/pubs/research_reports/RR630.html

Lorell, Mark A., Leslie Adrienne Payne, and Karishma R. Mehta, *Program Characteristics That Contribute to Cost Growth: A Comparison of Air Force Major Defense Acquisition Programs*, Santa Monica, Calif.: RAND Corporation, RR-1761-AF, 2017. As of July 23, 2020: https://www.rand.org/pubs/research_reports/RR1761.html

Mahnken, Thomas G., *Technology and the American Way of War Since 1945*, New York: Columbia University Press, 2008.

Mahnken, Thomas G., Grace B. Kim, and Adam Lemon, *Piercing the Fog of Peace: Developing Innovative Operational Concepts for a New Era*, Washington, D.C.: Center for Strategic and Budgetary Assessments, 2019.

Medeiros, Evan S., "The Changing Fundamentals of US-China Relations," *Washington Quarterly*, Vol. 42, No. 3, 2019, pp. 93–119.

Mizokami, Kyle, "In a Future War, the Air Force's Big Air Bases Could Be a Big Liability," *Popular Mechanics*, February 13, 2020. As of February 19, 2020:
https://www.popularmechanics.com/military/aviation/a30897139/air-force-air-base-strategy/

Office of the Secretary of Defense, *Nuclear Posture Review*, Washington, D.C.: U.S. Department of Defense, 2018.

Office of the Secretary of Defense, *Annual Report to Congress: Military and Security Developments Involving the People's Republic of China 2019*, Washington, D.C.: U.S. Department of Defense, May 2, 2019.

O'Leary, Lizzie, "The Modern Supply Chain Is Snapping," *The Atlantic*, March 19, 2020. As of September 19, 2020:
https://www.theatlantic.com/ideas/archive/2020/03/supply-chains-and-coronavirus/608329/

Pappalardo, Joe, "The Air Force Is Actually Considering Rocket Launches to Move Cargo Around the Globe," *Popular Mechanics*, October 29, 2018. As of February 19, 2020:
https://www.popularmechanics.com/space/rockets/a24406710/air-force-rocket-launch-cargo/

Public Law 114-92, *National Defense Authorization Act for Fiscal Year 2016*, November 25, 2015.

Ross, Robert S., "The Geography of the Peace: East Asia in the Twenty-First Century," *International Security*, Vol. 23, No. 4, Spring 1999, pp. 81–118.

Rovner, Joshua, "Two Kinds of Catastrophe: Nuclear Escalation and Protracted War in Asia," *Journal of Strategic Studies*, Vol. 40, No. 5, 2017, pp. 696–730.

Schlesinger, James R., *Systems Analysis and the Political Process*, Santa Monica, Calif.: RAND Corporation, P-3464, 1967. As of June 2, 2021:
https://www.rand.org/pubs/papers/P3464.html

Shlapak, David A., and Michael W. Johnson, *Reinforcing Deterrence on NATO's Eastern Flank: Wargaming the Defense of the Baltics*, Santa Monica, Calif.: RAND Corporation, RR-1253-A, 2016. As of June 2, 2021:
https://www.rand.org/pubs/research_reports/RR1253.html

Shlapak, David A., David T. Orletsky, Toy I. Reid, Murray Scot Tanner, and Barry Wilson, *A Question of Balance: Political Context and Military Aspects of the China-Taiwan Dispute*, Santa Monica, Calif.: RAND Corporation, MG-888-SRF, 2009. As of June 2, 2021:
https://www.rand.org/pubs/monographs/MG888.html

Snyder, Don, Kristin F. Lynch, Colby Peyton Steiner, John G. Drew, Myron Hura, Miriam E. Marlier, and Theo Milonopoulos, *Command and Control of U.S. Air Force Combat Support in a High-End Fight*, Santa Monica, Calif.: RAND Corporation, RR-A316-1, 2021. As of

October 20, 2021:
https://www.rand.org/pubs/research_reports/RRA316-1.html

Stanley, Elizabeth A., *Paths to Peace: Domestic Coalition Shifts, War Termination, and the Korean War*, Stanford, Calif.: Stanford University Press, 2009.

Suid, Lawrence H., *The Army's Nuclear Power Program: The Evolution of a Support Agency*, New York: Greenwood Press, 1990.

Tanenya, O. S., and V. N. Uryupin, "Certain Aspects of Employing the Airborne Forces in Russia's Arctic Zone," *Military Thought*, Vol. 8, No. 1, 2019, pp. 42–55.

Theis, Michael, Randall Trzeciak, Daniel Costa, Andrew Moore, Sarah Miller, Tracy Cassidy, and William Claycomb, *Common Sense Guide to Mitigating Insider Threats*, 6th ed., Pittsburgh, Pa.: Carnegie Mellon University, Software Engineering Institute Technical Note CMU/SEI-2018-TR-010, 2019.

Trimble, Steve, "Nuclear Air Force?" *Aviation Week & Space Technology*, November 25–December 8, 2019, pp. 52–54.

Trimble, Steve, "U.S. Air Force Launches Fielding Plan for Skyborg Weapons," *Aviation Week & Space Technology*, July 13–26, 2020, pp. 57–58.

Turner, Jobie, *Feeding Victory: Innovative Military Logistics from Lake George to Khe Sanh*, Lawrence, Kan.: University Press of Kansas, 2020.

United States Nuclear Regulatory Commission, "High-Level Waste Disposal: NRC's Yucca Mountain Licensing Activities," webpage, updated March 12, 2020. As of June 24, 2020: https://www.nrc.gov/waste/hlw-disposal.html

U.S. Air Force, *Air Force Glossary*, Maxwell Air Force Base, Ala.: Curtis E. LeMay Center for Doctrine Development and Education, July 18, 2017.

U.S. Air Force, *Agile Combat Employment for Force Providers, Annex A to Adaptive Operations in Contested Environments*, Washington, D.C., June 11, 2020, Not available to the general public.

U.S. Department of Defense, *Summary of the 2018 National Defense Strategy of the United States of America: Sharpening the American Military's Competitive Edge*, Washington, D.C., 2018.

Vaughan, Diane, *The Challenger Launch Decision: Risky Technology, Culture, and Deviance at NASA*, Chicago, Ill.: University of Chicago Press, 1996.

Waterman, Shaun, "Air Force Cyber Weapons Factory Moves Ahead with Blockchain Project," *Air Force Magazine*, July 17, 2020. As of August 31, 2020:

https://www.airforcemag.com/air-force-cyber-weapons-factory-moves-ahead-with-blockchain-project/

Wattenhofer, Roger, *Distributed Ledger Technology: The Science of the Blockchain*, 2nd ed., San Bernardino, Calif.: Inverted Forest Publishing, 2017.

Weitsman, Patricia A., *Waging War: Alliances, Coalitions, and Institutions of Interstate Violence*, Palo Alto, Calif.: Stanford University Press, 2013.

Westerlund, Fredrik, and Susanne Oxenstierna, eds., with Gudrun Persson, Jonas Kjellén, Johan Norberg, Jakob Hedenskog, Tomas Malmlöf, Martin Goliath, Johan Engvall, and Nils Dahlqvist, *Russian Military Capability in a Ten-Year Perspective—2019*, Swedish Defense Research Agency (FOI), FOI-R--4758--SE, December 2019.

Wright, David C., *A Dragon Eyes the Top of the World: Arctic Policy Debate and Discussion in China*, Newport, R.I.: U.S. Naval War College, China Maritime Studies Institute, August 2011.

Xiaosong, Shou, ed., *The Science of Military Strategy*, Beijing: Military Science Press, 2013.

Xie, Kevin, "Some BRICS in the Arctic: Developing Powers Look North," *Harvard International Review*, Vol. 36, No. 3, Spring 2015, pp. 60–63.

Yamada, Kei, "JAERO's Recent Public Opinion Survey on Nuclear Energy: Support Rises Somewhat for Restarting NPPs," Japan Atomic Industrial Forum, Inc., webpage, March 22, 2019. As of April 8, 2020:
https://www.jaif.or.jp/en/jaeros-recent-public-opinion-survey-on-nuclear-energy-support-rises-somewhat-for-restarting-npps/